Kenzhebek Badanov

Fabric printing technology

Kenzhebek Badanov

Fabric printing technology

ScienciaScripts

Imprint

Any brand names and product names mentioned in this book are subject to trademark, brand or patent protection and are trademarks or registered trademarks of their respective holders. The use of brand names, product names, common names, trade names, product descriptions etc. even without a particular marking in this work is in no way to be construed to mean that such names may be regarded as unrestricted in respect of trademark and brand protection legislation and could thus be used by anyone.

Cover image: www.ingimage.com

This book is a translation from the original published under ISBN 978-620-6-79246-8.

Publisher:
Sciencia Scripts
is a trademark of
Dodo Books Indian Ocean Ltd. and OmniScriptum S.R.L publishing group

120 High Road, East Finchley, London, N2 9ED, United Kingdom
Str. Armeneasca 28/1, office 1, Chisinau MD-2012, Republic of Moldova, Europe
Printed at: see last page
ISBN: 978-620-6-68415-2

K.I. Badanov

Fabric printing technology

Study guide

Taraz 2020

UDC 677.02 (075)
BBK

Badanov K.I.

Technology of fabric printing. Taraz: Taraz University, 2020. - 129 c.

ISBN

Reviewers:
J.S. Rakhmanova - Candidate of Technical Sciences, Professor of Taraz Innovation and Humanitarian University.
B.M. Ualiev - Candidate of Technical Sciences, Associate Professor of the Department of "Design and Technology of Garments" of Taraz State University named after M.H. Dulati.

The physical and chemical essence of printing processes, printing methods, fabric printing technology are considered. Along with classical methods are given new ways of printing with the use of computer technology in printing. The physical and chemical essence of the technology of printing flat and rotary mesh, as well as the operation of equipment for dye fixation is considered in detail. Information on printing with pigments is given. The technology of direct printing in laboratory conditions is given.

The textbook is intended for students, masters and PhD doctoral students of specialty 050733-Technology and design of textile materials, as well as specialty 050726-Technology and design of light industry products.

BBK

Approved for publication by the Academic Council of Taraz State University named after M.H. Dulati (Minutes №_10_ from "_16__" _06____ 2020).

Table of Contents

Introduction

Textile industry of the Republic of Kazakhstan is a group of enterprises engaged in processing natural and synthetic fibers into yarn and fabrics. According to the General Classifier of Economic Activities of the Republic of Kazakhstan (OKED), textile production should be represented by 7 types and 20 subspecies of economic activities. Due to the decline of the industry, many production directions have ceased to exist, closed. For example, there are no such activities as production of sewing threads, dyeing of fibers and yarns, bleaching, dyeing, filling and finishing of fabrics, production of knitted fabric and others.

At the present stage, the share of textile and clothing industry in the total gross output of the country is 0.4%, in the volume of industrial production - 1.3%.

In developed countries, such as Germany, France and the United States of America (hereinafter - the USA), the share of textile and light industry in the volume of industrial production is 6 - 8 %, in Italy - 12 %. This allows them to form 20% of the budget, as well as to ensure the filling of the domestic market by 75 - 85% with products of their own production.

The textile and clothing industry of the Republic of Kazakhstan covers only 10% of the needs of the domestic market. While for the formation of economic security of the country the volume of domestic production should at least satisfy 30% of domestic demand. According to the Association of Light Industry Enterprises of Kazakhstan, about 80% of the market is made up of illegally imported imports. The problem of import substitution in the light industry of Kazakhstan can be solved only through a significant increase in competitiveness.

Kazakhstan has a good scale of market opportunities, both for the development of textile industry and separately taken sector of cotton-textile industry of the region. The sector of enterprises, which carried out finishing of fabrics, is poorly developed. Nevertheless, with the development of cotton fabrics production, this segment of the cotton-textile industry of the country will be restored. According to preliminary estimates, this segment can be quite profitable and does not require such large investments in equipment. The development of this segment is important to create a more competitive textile industry. The textile industry in Sri Lanka has made great strides by developing this particular segment of the production chain. Lacking a well-developed textile industry, Sri Lanka has concentrated cluster building on dyeing, printing and finishing. By achieving excellence in quality, design and speed in dyeing, printing and finishing, Sri Lanka's textile cluster has become globally competitive. The experience of Sri Lanka and other countries that applied this way of cluster development will be applied in the process of selection of investment projects and construction of textile industry enterprises on the territory of FEZ "Ontustik".

The most urgent problem for Kazakhstan's light industry is technical and technological backwardness of the industry. Many enterprises lack modern methods of coloring, final processing of fabrics that affect the properties of products. Competitiveness of the textile industry strongly depends on the level of investment in technology and innovation. It is expedient to introduce automated control systems, as well as the use of computer programs for design and modeling, allowing to reduce to the maximum extent possible the work on graphic design, long and labor-intensive sketching.

1 Printing processes

Printing on cloth was earlier than printing on paper. The simplest methods seem to have been known in the East: for coloring, part of the cloth was covered with wax, or places that were to remain unpainted were bound. The printing technique comes from the Copts, from Egypt. Originally it was monochrome, more often blue (indigo was used). In the Middle Ages, fabrics imitating more expensive materials (brocade, velvet) appeared. The two oldest printing methods are batik and silk-screen printing.

Batik is a hand painting. The first mentions of painting fabrics with wax patterns are found already in Pliny's "Natural History" and in Chinese manuscripts of the VIII century. There are three main types of batik:
- cold;
- hot;
- free painting.

The essence of the technology itself is the same - the areas of the fabric not subject to dyeing are covered with various resins or beeswax and, absorbed into the fabric, protect it from the dye. The prepared fabric is dipped into the paint, then remove the wax composition and the result is a white drawing on a painted background. Then the artist paints the product with brushes. Here the result depends only on his skill. The product is steamed, cleaned of excess paint, dried and ironed. The manufacturing process is very labor-intensive, so the cost is very high.

Silk-screening - the emergence of silk-screening as an art and craft dates back to the Qin Dynasty in ancient China, the only producer of high quality silk at that time. The root "silk" in the word "silkscreen" means that the ink was pressed through a silk mesh when the image was applied. Silk, as paradoxical as it may sound, is no longer used in silkscreen printing. In the modern sense, this method of printing was patented in 1907 by a Simon from Manchester (in England) under the name SILK SCREEN PRINTING ("printing through a silk screen"). The rapid introduction of this method into industry began during World War I, when chain stores began to appear in the United States. Each chain had to have its own corporate style - signs, shop windows, aprons and hats with corporate symbols. But all this was required in small print runs - 50, 200, at most 1000 pieces. The method improved and captured more and more new areas of application. Firstly, it allows printing on almost any material - from balloons and plastic bags to rocket bodies. Secondly, it perfectly combines and complements any traditional products, decorating them. Thirdly, it allows you to make images of any size - from miniatures to huge canvases. In order to print one color you need to make one stencil on a special frame. If there are 6 colors, then six frames are prepared.

Stuffing was done with wooden molds. In modern times, the film method was used to create more complex ornaments, but wooden molds were used when folk or folk-inspired fabrics became fashionable.

Printing of fabrics is a method of their artistic and coloristic design, which differs from smooth dyeing by the fact that dyeing takes place only in places where a pattern consisting of one or more colors is applied. Therefore, its realization requires more sophisticated equipment and additional fabric treatments.

Printing produces a colored pattern on fabrics or garments. Each color is applied separately. The color of the printed pattern can be created either by pigment or dye. The pigment does not penetrate the fibers, but is fixed to the surface of the fabric by a binder. The process of pigment printing is relatively simple, as no washing of the fabric is required at the end of the process. One of the disadvantages of pigment printing is that the desired "fabric quality to the touch" is not always achieved. Improvements in fabric quality and pattern stability with modern binders have increased the popularity of pigment printing. Dye printing is a more complex process, as the dye must be fixed in the mills by steam or heat, and then the fabric must be washed to remove the loose dye. Among the advantages of dye printing are the softness of the fabric, the brightness of the colors and the good stability of the pattern.

The pattern is printed on the fabric using fabric printing machines. In order to fix the printing ink, the fabric is usually subjected to steam treatment after drying. In the steam environment, the dye passes from the surface into the fabric and individual fibers. To remove the thickener and the dye deposited on the surface of the fabric, it is washed. If necessary, special treatments, e.g. oxidizing agents, are carried out on the washing machines to finally fix the dye to the fabric.

In addition to printing fabrics, colors are sometimes produced on carding sliver ("figure") and yarns, both in skein and warp form ("flamme"). In some cases, patterns on fabrics are obtained by gluing (in an electrostatic field) to the surface of the textile material finely cut fiber-floc. For special purposes photochemical method of printing is used.

2 Basic information about batik

There are several ways to decorate the material, one of the most popular is batik. This name means hand painting on fabric with the use of so-called reserving compositions. Drawings, created by your own hands, turn out not only beautiful, but also spectacular.

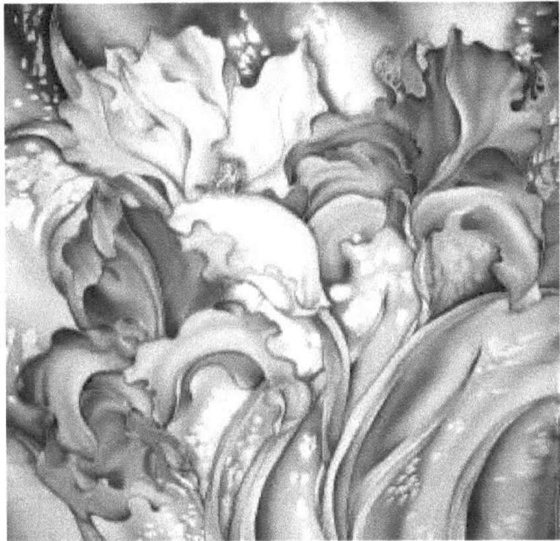

Figure 1: Hand painting on fabric - batik

The term batik is a generalized name for various methods of hand-painting fabric. The word "batik" is translated from the Indonesian language as "drop of wax". The specificity of such painting is that on the fabric - cotton, silk, synthetic or wool - is applied a coloring substance using wax or gasoline to prevent its spreading and to mark the boundaries of the pattern. This technique is based on the principle of reserving, i.e. covering with a special composition those areas of the material that should remain unpainted, so that in the end the desired pattern is obtained. To obtain clear contours (boundaries), a special fixer called reserve is used, which is made on the basis of gasoline, paraffin, water base.

The technique is based on the fact that rubber glue, paraffin and certain types of resins and varnishes that are applied to the material and do not allow the dye to pass through. As for the fabrics, for the painting of which the batik technique is used, they are cotton, silk, wool and synthetics. There are several types of batik for getting a garment with a painted surface, for example, a T-shirt.

The island of Java (Indonesia) is considered to be the birthplace of hand painting on fabric with the use of reserve. In translation from Javanese batik technique means "drop of wax". This type of painting has long been known among the peoples of modern Indonesia and India. In general, the mankind is engaged in creating a pattern on the fabric since time immemorial. Different ways of painting fabrics were known in Sumer, Japan, Peru, Indochina, Sri Lanka and some African countries.

Figure 2: Hand-painted fabric on Chinese silk

The birthplace of batik in the modern understanding of this technique is Java. The masters of this Indonesian island were able to reach unprecedented heights and sophistication in terms of dyeing fabrics. They managed to make a whole art out of this craft in a short period of time. It could take months, and even years for the masters to create one cloth. This is due to the multi-stage preparation of the fabric: soaking, boiling, bleaching, a long process of dyeing with the application of wax, dyeing, drying. Because of the lengthy process, originally only aristocrats wore clothes that were covered with patterns using batik. The craftsmen of Java used cotton fabrics, passing the secrets of the craft from generation to generation. An interesting feature is that the patterns of different families could differ significantly from each other. The patterns made had a variety of themes, from plant ornaments and geometric patterns to mythological subjects. On the northern coast of the island lighter shades are used than in the central part of the island. By the patterns on the clothes, it was possible to determine to which caste a person belonged. It was forbidden to repeat traditional royal ornaments. Every girl in her dowry must have things decorated with batik technique. For example, colorful fabrics, curtains, wall panels, closet items. Even today, hand-painted clothes are in great demand among the peoples of Java, and are often made for commercial purposes. In the XVIIth century, when Java became a Dutch colony, batik began to arrive in the countries of Europe. Later, Europeans invented the electric batik pin device, which could be used to store wax in a molten state.

In 1801 José-Marie Jacquart created an automatic machine with special punch cards, thanks to which it became possible to produce canvases with drawings created by artists. Thanks to this batik technique moved to a new level. It became most popular in Europe only at the end of the XIX century.

The technique of painting fabric is divided into several types, depending on what materials are used, and what stages must be overcome to obtain a surface with a spectacular pattern. The peculiarity of this or that type is that one option is ideal for synthetic fibers, the second - for decorating silk, etc.

Hot batik. Wax is used as a reserve. A special tool called chanting is used to apply it. The wax helps to limit the spread of the dye as it does not absorb it. This type of painting is called hot painting due to the fact that the wax used in it is necessarily melted. The paint is applied in several layers. At the end of the work, the wax is carefully removed. This method is used for painting cotton fabric.

Cold look. Ideal for decorating artificial materials, silk. The technique uses dyes made on the basis of aniline. Reserve is thick, if it has rubber components and not thick, when the basis is gasoline. Rubber ones are applied from tubes, and gasoline ones with the help of glass tubes. In addition, both colored and colorless reserves can be used. Cold type implies a single-layer application of paint. The work requires more accuracy from the performer compared to hot technology.

Free painting. It is widely used on materials made of natural silk and synthetic fibers. For it masters often use oil paints and aniline dyes.

Folded batik "sibori". The peculiarity of this type is that the master performs tying the cloth in a certain way, and only then dyes it. *Knotty kind.* In this case, the dyed fabric first make a lot of small knots, tying each of them with a thread. After the surface is dyed, they are carefully removed.

Figure 4: Bandan or knotting technique

3 Technique of batik realization

Paints are applied to the canvas so that clear and distinct boundaries are obtained at the junction of different shades. For this purpose, a reserve is used, i.e. a special fixer based on gasoline, paraffin, etc. - The composition varies depending on the chosen technique, material, colors. Basically, the following types of techniques are distinguished: cold; hot; free painting; free painting with salt solution.

Cold batik Painting on batik fabric as a hobby is suitable mainly for patient people, as this process is labor-intensive and time-consuming. One of the popular techniques is cold, which appeared much later than hot, with the development of the chemical industry. Its appearance simplified the work. The role of wax in the cold technique is performed by special reserves, which do not need to be heated, etched and reapplied. To work, it will be necessary to create a separate workplace. The ideal option is a well-ventilated room. This is due to the fact that the vapors of the reserving substance have not exactly a good effect on health. If you are going to use the painted material for use as a tablecloth, scarf, etc., then take into account that you need to fix the paint: baking in the oven, steaming on a water bath without contact with condensate / water, iron. If this is not done, the first wash will wash off all the work. You will need: a simple pencil; reserve (black), a glass tube for it; kalanok brushes, aniline compositions; buttons, stretcher; natural silk (crepe de chine). Decide on the choice of sketch. For it, you will need a sheet of thin paper. If you want something spectacular, then give preference to colors. Applying elements on the canvas, try to draw them so that each of them has a closed contour. Apply the reserve on the contours should be without delay, but also without haste. Sequence of actions: First, wash the material, then stretch the pre-dried canvas on the stretcher, using buttons. Take a glass tube, fill it with reserve. Apply the composition to the contours of the elements. To increase the number of shades, dilute the same paint with different amounts of water. Use yogurt jars, disposable cups. Paint the colors (from light to dark tones) and the background. Take salt, sprinkle it on the canvas and let it dry. Shake off the salt and once the cloth is dry, remove it from the stretcher. After 24 hours, boil the cloth (about 3 hours), wash it in warm soapy water. Be sure to rinse the creation, adding a little vinegar to the water. Gently wring out the product and iron it while it is damp. After the end of the procedure, blow the reserve back into the container, and rinse the glass tube in gasoline. Otherwise, the residue will harden and the tool will become unsuitable for further use.

Hot batik on clothes looks beautiful and spectacular. It is suitable for those who do not like to painstakingly paint each piece of cloth, sitting for this work for several hours. Even without much effort you can get a canvas, which will make exclusive skirts, scarves and even suits. Work with this technique is carried out with melted wax, stearin, paraffin or their mixture, so be careful. List of tools that may be needed when decorating the cloth: natural fabric, for example, cotton, wool, silk, cardboard stencil, dyes for painting on fabric, a glass for water, wax, chanting, brushes, rubber gloves, cellophane, newspapers: hair dryer or iron. For work, it is better to wear clothes, which will not be sorry to spoil, because the paint for the fabric is practically not washable. As an option, put on a waterproof apron. Hot batik technology consists of the following actions: To apply one of the above solutions to the canvas, use a special tool - chanting. It looks like a watering can with a thin tip. Although, recently, brushes have become widely used, with the help of which strokes and dot drops are applied to the fabric. After that, you need to apply a layer of paint on top. Next, you can again apply wax and another type of paint to some areas. To make the patterns become organized, use stamps that need to be dipped in melted wax. You can use 2-3 tones or more. Once the paint is dry, get rid of the wax. For this purpose, put a newspaper on the cloth and iron it - it should absorb the melted substance. Then put a new newspaper and repeat the procedure. Do this until there is no wax left on the cloth at all.

Free painting. Thanks to this technique of batik you can show all your abilities in the field of drawing, because here you can not form a drawing on a certain template. With the help of free painting creates an individual and unique work. Basically, this type of technique is practiced with the use of oil paints with special solvents or aniline dyes. You can even make experiments, adding somewhere salt effect or reserving composition, or using alcohol-containing substances to moisturize the fabric. Ways of freehand painting in batik: with paints, thickening from the reserving liquid; with paints, saline solution; printing paints; oil paints; on silk, applying thickening agents.

Free painting with the use of salt solution. The essence of this technique is that the fabric stretched on the frame, depending on the particular pattern, impregnated with an aqueous solution of salt (table salt) and after it dries, the painting of the cloth is made. In some cases, it is carried out by colors from the basic dyes, in which was introduced salt solution. This approach helps to limit the spreadability of paint on the fabric and provides an opportunity to create drawings with free strokes. In this case, you can vary both the shape and the degree of saturation of one or another color. It should be added that free painting with paints in which the composition of the solution of table salt, can be successfully combined with conventional painting cold batik. For this purpose, some fragments of the pattern are created by free painting with finalization of the graphic pattern. Background overlaps in this case are carried out in areas that are limited by the reserve. Instead of salt, gelatin or starch can be used as a ground. Deciding to delve into this craft and try to create a spectacular drawing in this way, prepare the following tools, materials: aniline dyes: salt solution; a cut of silk fabric, which is stretched on the frame; sketch of the future work; brushes for working with paints of different thicknesses, flute brushes for impregnating the material, synthetics; pipette for picking up paint, water for washing brushes; palette; soft pencil 8B. To prepare a salt solution, take a couple tablespoons of salt per glass of water. Dilute the salt in hot water in an enameled or glass bowl so that all the crystals dissolve. Depending on the task at hand, make a preliminary painting with a very soft pencil on stretched silk. Salt painting consists of three steps: *Raw painting*. Prepare the desired colors on the palette before the silk is soaked in the salt solution. Also, use only the salt solution to dilute the colors. While the silk is still damp, use broad touches or brush strokes to apply the colors in their proper places. On very wet fabric, the paint will begin to spread in unexpected shapes. Painting on semi-dry silk and applying secondary spots. As long as the silk is damp and salt crystals form on it, you can achieve painterly effects. The paint won't run uncontrollably in all directions, but it will leave a brush mark that will blur a bit later on. This is especially ideal for imitating natural textures. Graphic drawing of details related to the first plan. As soon as the salt solution dries, the fabric will become crisp, and the paint will barely noticeably spread over it. Continue the work by drawing the details by applying short strokes or putting different sized spots. Batik for beginners Engaged in painting fabric, you will definitely feel your involvement in high art, especially when you start to get spectacular patterns and designs on the fabric. Over time, you will be able to develop your own design and style, in which different items of your closet will be kept. In the beginning it is better to stop on the technique of cold batik, because the hot variant is more complicated, labor-intensive. Strictly adhere to the technology, so that the made work turned out the necessary quality: At first it will be necessary to stretch the fabric on the frame so that it does not sag. Batik masters often stretch wet cloth, since after drying it is made more elastic. Prepare a sketch, which is drawn with a pencil on paper. Next, the sketch is transferred to the fabric. For this purpose, the paper is fixed under the fabric so that the lines of the pattern can be traced on the cloth with a soft pencil. Trace the outlines using a reserving compound. To dial the reserve, lower the spout of the glass tube into the container with the contouring liquid, and in the other end insert a syringe (rubber) - with its help the liquid will be sucked into the tube. Then check the circuit. After complete drying of the reserving composition you need to pass a brush soaked in water over the entire drawing on one side of the contour, and after a while to make sure that the water has not gone over the reserve line. If you find places where the contour will be weak, then after complete drying of the matter again pass the reserve over these areas. On the sixth step, take up the coloring of the painting. Be extremely careful. In the end, remove the work from the frame and fix the batik by ironing, baking or as an option, you can steam (steam) on a water bath.

What you need to prepare for the first lessons. Before you begin to master one of the batik techniques, prepare the appropriate accessories. At the same time, remember that the success of fixing the colors also depends on the quality of the paints used. In any case, wash clothes with batik should be washed in cool water with the addition of a small amount of vinegar. For the first lessons you will definitely need: hoops or frame. If you plan to apply the painting to a small area, then give preference to embroidery hoops. If you are going to work with a large format, then prepare a special frame for batik - an ordinary stretcher will do. Fix the fabric on the frame should be on the hooks that come with it in the set. As for the stretcher, you can fix the fabric on it with a thread and a needle, and do it in such a way that the fabric does not touch the frame. A simpler option is to fix the fabric with a furniture stapler or buttons.

Paper. Prepare a sheet of thin paper to make a preliminary sketch. The size of the sheet should be equal to the area of the drawing on the canvas.

Fabric. It is best to use thin natural fabrics such as silk, batiste, double-filament. Dense material is not suitable, because the reserving composition may not pass through thick fibers, resulting in an "explosion" of paint - one color begins to move to the edges and zone of the other. Beginners are recommended to choose batiste.

Materials, tools. In a standard batik kit there is a reserve, a glass tube for it and paints. In some sets there is a contouring liquid, ready for application - it is poured into a tube with a thin tip. You can weld the reserve yourself, but this activity is not only labor-intensive, but also fire hazardous. Drawings for batik for beginners Relatively simple options for beginners is a drawing in the form of a composition of flowers. It looks great on children's and adult clothes, canvases that decorate rooms. To paint a whole bouquet on the matter, perform the following actions: draw three ovals of different sizes, in the center of each oval depict a wavy flower core, and at the bottom - the stem, draw around each core a flower, depict the top of the right side of the flowers bud, make more voluminous stems, podrisuite to each of them leaves, draw leaves and around the flowers, at the end of the neatly erase all the auxiliary circles. Another equally interesting and quick option is a bouquet of roses. To get such a sketch, draw several circles on paper, then transform each of them into a blossoming layered bud. Practice on paper, so that in the future you will be able to draw roses by reserve on matter the first time. Draw bouquets of flowers on the canvas will have to draw without any auxiliary lines.

4 Methods of printing textile materials

Printing is a patterned coloring of fabric obtained on its surface by thickened or liquid printing ink with subsequent fixation of the dye on the fiber. Printing of fabrics can be carried out: by hand stuffing; by airbrushing; by cylindrical fabric printing machines with engraved rollers; by flat or cylindrical mesh templates; by polychromatic patterned dyeing, by transfer printing, etc.

Hand stuffing is done with hand molds called manners or blooms, which are a plate with a relief pattern cut or typed on its surface.

Figure 5: General view of the manner

The essence of this method of artistic design of fabric was as follows: a pattern was cut out on a board, then the board was moistened with paint, turning into a printing plate (manner). The inked form was put on the fabric spread out on a table covered with cloth and lining, then hit the handle of the form with a heavy hammer. Hence the name of the method "stuffing".

Figure 6. Hand-stamping of the pattern with manners

Wood block printing with a pattern protruding from the surface was practiced hundreds of years ago. Although wood block printing for textile printing is now a craftsman's domain, templates with protruding porous surfaces are used for carpet printing.

Printing on textiles as such originated in the XVIII century, when the first devices designed to automate this process appeared. For this purpose, rollers made of various materials with engraved designs were used.

Figure 7. Shaft with protruding surfaces

Flexographic printing is also used, where the relief pattern is created on the rubber coating of the printing roller. Flexographic printing is also used in the printing industry.

Cubicle padding.

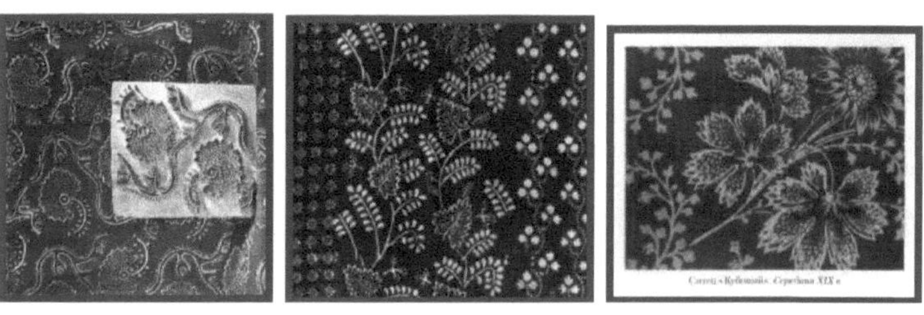

Pavlovo-Posad naboyka

Of great interest is a unique collection of printed fabrics by the Swiss collector and scientist Robert Forrer. It is now located in the famous safe room of the Russian State Library. Until that time the collection, including 238 samples of XIII-XVII century fabrics, as well as hand-printed or printed fabrics of the XVIII and XIX centuries, was an active exposition of the German Museum of Book and Type (Leipzig)

5 Fabric printing technology

In order to obtain high quality printing with a clear outline, the printing ink must have the following properties:
- do not spread when applied to textile material;
- be homogeneous in composition, penetrate easily through the template without clogging it, or easily fill the print roller engraving and exit it during the process of drawing on the product;
- be sticky, viscous, yet easily wettable and rinse off well when rinsed;
- swell quickly and evenly when steaming and do not prevent the dye from transferring to the fiber.

Most of the above properties of the printing ink are due to the presence of a thickener. One of the main requirements for the thickener is that it must not interact with the dye and textile auxiliaries in the printing ink.

The technology of patterned coloring consists of the following operations:

Applying the printed pattern to the fabric → drying → ripping → washing.

A necessary condition for obtaining high-quality printed fabrics is their good preliminary preparation for printing by chemical and mechanical treatments. The better chemical preparation of the fabric (raschlichtovka, boiling, bleaching), the higher its capillarity and better absorption of printing ink by fiber, hence, brighter coloring of the pattern. Mechanical preparation of fabrics consists in their shearing and cleaning from fluff, smoothing to remove folds, serrations and creases, in giving standard width and eliminating warps of weft threads.

Patterning on fabric can be done in a variety of ways

in a number of ways:
- by cylindrical tissue printing machines using engraved rollers;
- with grid patterns;
- with a heat transfer printing process;
- non-traditional methods (hand-printing, polychromatic dyeing, airbrushing, etc.).

Printing on machines with engraved rolls is used for cotton, viscose and mixed fabrics, V=80 m/min.

Printing on machines with flat mesh templates (photofilm printing), used for silk fabrics and knitted fabrics. Patterns of any complexity are obtained. The disadvantage is low productivity.

Printed on machines with cylindrical (rotary) mesh templates. Applicable for all fabrics and knitted fabrics. Modern high-performance machines, V = 100-150 m/min.

Printing on thermal presses or thermal calendars (transfer printing).

The choice of printing method is determined by the physical and mechanical properties of the fabric and the nature of the pattern. The most important operation of fabric processing after printing is the fixation of dye on the fabric under the action of moisture and heat. It is carried out in special chambers called ripening chambers.

Drying machines for printed fabrics have their own design features and are aggregated with printing machines. Washing of fabrics after printing is no less important technological operation than the processes of printing itself and fixation of dyes on the textile material. The quality of washing determines the operational and hygienic properties of printed fabrics, coloristic indicators of patterned colors, significantly affects the economic characteristics of the technological process.

Cylindrical printing with engraved rollers, grid patterns and transfer printing have gained industrial importance. The use of polychromatic patterned coloring methods is still limited.

According to the method of creating patterns on textile material, a distinction is made between direct, etched and reserve printing.

Direct printing is the application of a pattern on white or lightly colored fabric. It is obtained either white-ground fabrics with a small area occupied by drawings, or ground fabrics, when most or all of the surface is occupied by printed designs. Varieties of direct printing are: raster printing, i.e. the application of designs with three dyes (bright blue, bright red and bright yellow), which are applied to the fabric in pure form or mixed.

Technological scheme of direct printing:

Printing on white or lightly dyed fabric and drying on the printing press → steaming or heat treatment to diffuse the dye into the fiber → washing to remove thickening and loose dye → drying.

Etching printing - printing on dyed fabric by etching (oxidizing agents, reducing agents, etc.) that discolors the dye. During the subsequent treatment of the fabric with hot steam, the dye is destroyed and white patterns appear on the dyed fabric. If at the same time with etching a new dye of a different composition is applied to the fabric, colored patterns may appear. This method of printing on fabrics is often used for making coats of arms and flags.

Technological scheme of etched printing:

Dyeing of fabric in dark color → washing → drying → printing with etched printing ink → steaming to break down the dye (8-10 minutes) → washing → drying

Reserve printing - printing on white fabric with reserve (wax, stearin, salts, reducing agents, etc.) followed by dyeing with dye solution. Reserved areas are not colored and after removal of the reserve there are white patterns on the dyed fabric. The dye is fixed on the entire area of the fabric except for those areas on which the reserve composition is applied, preventing the dye from fixing. Colored or white patterns can be produced, depending on whether the reserving composition contains or does not contain a dye that is resistant to the dye.

Technological scheme of backup printing:

Printing with reserve printing ink and drying → heat treatment to fix ink components → plating with dye solution → steaming to diffuse dye into fiber → washing → drying.

Etching and reserve printing methods have practically lost their significance at present, and there are no reports in the literature about the development or improvement of these methods. Their application is economically feasible only for obtaining small, sparsely scattered patterns on the surface of the fabric, and such coloristic effects, if necessary, can be obtained by a simpler method of direct printing with appropriate pigment compositions on dyed textile materials.

6 Printing fabrics on machines with engraved rollers

These machines are used for printing cotton and viscose staple fabrics. Printing is accomplished by the ink from the recess of the engraving onto the fabric as it passes between the roll and the truck. The printing machines are designed to produce patterned fabric colors by means of rotating engraved cylindrical rollers. The machines are designed for high-performance printing of fabrics with thickened compositions of dyes of various classes by means of cylindrical shafts with deep engraved engravings. There are single-shaft and multi-shaft printing machines. Multi-shaft have an even number of rollers, but no more than 16. The main working organ is rotating cylindrical engraved rollers with deepened engraving. The advantages of fabric printing machines include their high productivity, reliability, the possibility of obtaining complex multicolor patterns. The disadvantages are high fabric tension, difficulty of readjustment when switching to a new pattern, technical complexity of design, significant dimensions, energy consumption.

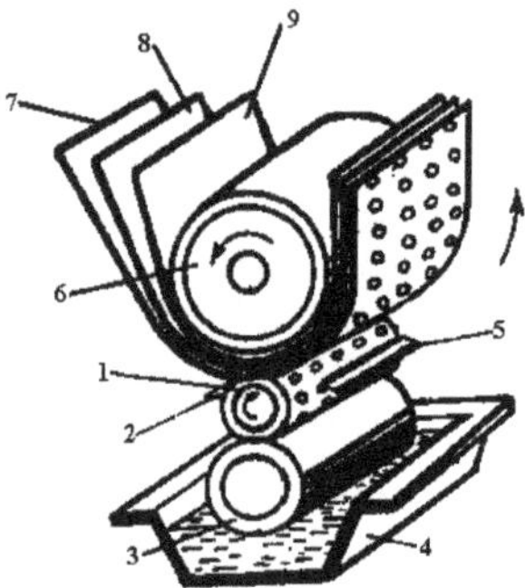

1- engraved roll; 2 - counter-rake; 3 - ink brush; 4 - ink trough; 5 - rakla; 6 - truck; 7 - printed fabric; 8 - cover; 9 - lapping

Figure 8. Printing press with engraved rollers

Single-shaft machines for one-color (monochrome) patterns and multi-shaft machines up to 16 rolls (polychrome patterns) are used.

a)   б)

Figure 9. Monochrome (a) and polychrome (b) printing

These machines are highly productive, with fabric speeds of 120-130 m/min or more. The printing rollers can have relief and recessed engravings, from which the ink is transferred to the fabric. For printing fabrics, machines with rollers having deepened engravings are used.

Printing presses come in single and multi-shaft versions. The most popular are eight-shaft machines. Work on them can be carried out with or without a cover. Coverless machines, in their turn, can be with a bale washer or with a rubberized truck. The use of a cover is necessary when printing multi-shaft designs with complex etching and for ground printing on thin fabrics that easily allow ink to pass through. Fabrics made of synthetic fibers can be used as covers, which, due to their high strength, hydrophobicity and low susceptibility to dyes, can withstand up to 1400 passes on printing machines. Coverless printing is more cost-effective and productive. Even more economical is the method of printing on machines with a rubberized truck. However, its disadvantage is that around the truck instead of printing rollers have to place washing devices that take up a lot of space because of this reduces the number of colors when printing.

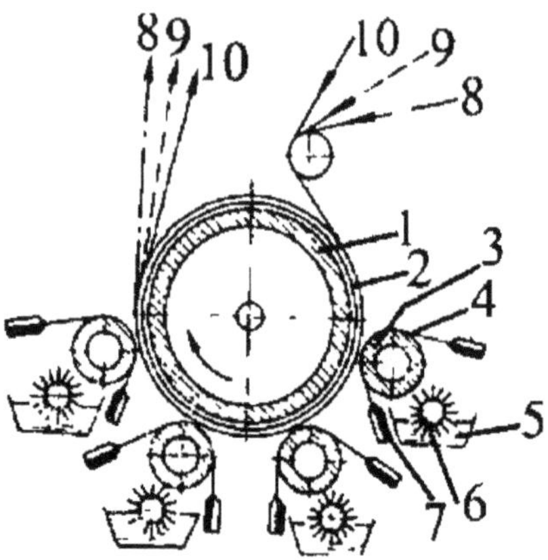

1- truck; 2 - lapping; 3 - printing roll; 4 - rakla; 5 - trough; 6 - brush; 7 - counter-rakla; 8 - cloth; 9 - cover; 10 - pickle

Figure 10. Four-shaft printing press

The printing rollers 3 are arranged around a central cylinder 1, called a truck or press, and are pressed with considerable force to its elastic-elastic surface formed by a multilayer coating 2 made of a special fabric called lapping. The *truck* has a diameter depending on the number of shafts. The truck rotates due to frictional forces. It is covered with laping on top to create the necessary elasticity of the truck. This makes it easier to remove the ink from the engraving and transfer it to the fabric. The elastic radial deformation of the laping in the area of contact with the printing roll improves the conditions for transferring the ink from the engraving to the fabric. The ink is applied to the printing roll with the help of a rotating brush 6 installed in a special printing machine trough-chassis 5, into which the pump pours the ink. To the printing shaft in the course of its rotation is pressed sharply sharpened thin steel plate-raklya 4, which cleans the surface of the printing rollers from excess ink, leaving it only in the recesses of the engraving. The rakel plate is usually 70 mm wide and 0.35 to 1.0 mm thick. With the help of a separate electric drive rakla reported to the reciprocating motion along the axis of the printing shaft, which increases the efficiency of cleaning the surface of the shaft and reduces the risk of mechanical damage to its engraved surface and the blade of the rakla. On the opposite side of the shaft against the course of rotation placed another plate, called counter-rakley 7. It is made of soft metal (brass) and serves to clean the printing roller from the fluff and ink, brought by the fabric. On multi-shaft machines counterrakli put usually only at the front printing rollers. The fabric to be processed 8 is fed into the machine and placed on the cover 9. Under the cover is an endless web 10 (kirza). The *kirza* is an endless cloth made of special cloth or rubber, designed to prevent pollution of the truck with paint. The kirza consists of 5-6 layers of durable cotton fabric 1.5-5 mm thick or more, rubberized on the front side. Kirza, wrapping around the truck, passes an endless ribbon through the entire machine at a significant tension, which is created by a special tension roller. If the printing unit includes a kirz washer, then furless printing is carried out. The *kirzo washer* washes, dries, cools and cools the kirzo for cuffless printing.

Otherwise, a *cover is* passed between the kirsa and the fabric to protect the kirsa from contamination. With the help of a truck, a flat, elastic and flexible surface is formed under the printing roll, i.e. a kind of "printing table". The rollers are pressed with great force against the truck and against the continuously moving fabric, and the elastic printing table facilitates the extraction of ink from the engraving onto the fabric. In this way, conditions are created for the continuous and successive application of a large number of printing inks to the fabric, applied accordingly to the pattern. The cover should be wider than the pickguard to protect it, the lapping and the truck from contamination by the printing ink.

For a better understanding of the interaction of the working parts, a schematic diagram of the main working unit of the printing press, called the printing bed, is given.

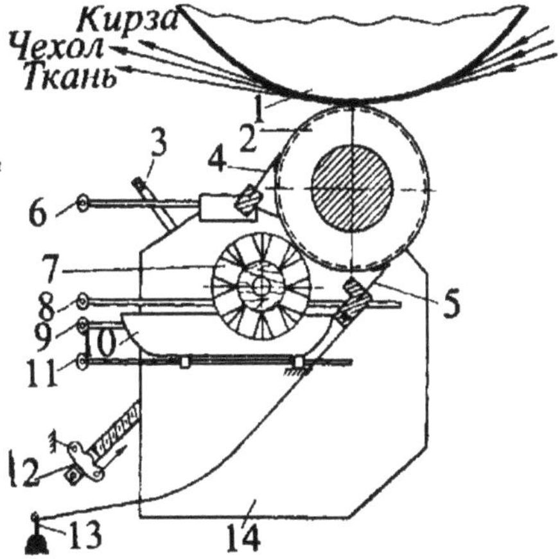

Figure 11. Schematic of the printing area

The printing roll 2 is placed on movable bearings between two planes 14 (jaws) and is pressed by the stop screw 12 against the truck 1. Receiving rotation from the drive, it provides transportation of fabric, cover and pickle and rotates the truck, which has no separate drive. A rotating brush 7, mounted in a trough 10, is pressed against the printing shaft. As the brush rotates, the ink from the trough is applied to the printing cylinder in a continuous layer. Excess ink is removed from the surface of the cylinder rakley 5, pressed against it with a lever with a weight 13. The ink remaining in the engraving (shown with a dotted line) is extracted with a cloth, and particles of fluff adhering to the cylinder are removed by counter-rake 4. The mutual position of the mechanisms is regulated by turning the set screws for moving the trough chassis 11, setting the rakli 8, moving the brush 9 and setting the counter-rakli 6. Screw 3 makes it possible to eliminate shaft misalignment, but in modern machines shaft misalignment is eliminated.

For drying the fabric and cover, the printing presses are combined with drying machines. For drying the fabric and cover, the printing presses are additionally combined with kirsa washers. They are used for washing and drying the kirsa. The printing rollers are hollow thick-walled cylinders made of red copper. They lend themselves well to machining and etching. Since copper is expensive and scarce, you can use copper-plated

shafts steel seamless pipes, on the surface of which galvanically build up a layer of copper 1-10 mm thick. Another method is to press a copper jacket 10-15 mm thick onto the shafts. A special press is used to press the shaft onto a steel spike, which serves as the shaft axis. On some machines printing shafts on the spike is not put on the spike, and installed on the cone supports, which are fixed shells of the printing shaft. The circumference lengths of shafts of the same pattern must be strictly identical. Deviations of no more than ± 0.2 mm are permitted. Irrespective of the size of the pattern, it must be laid out an integral number of times along the circumference length. The working surface of the printing roll consists of blank and printed elements. The former are a smooth, well-polished cylindrical surface, the latter a deepened engraving capable of holding the ink and giving it to the fabric. There are many ways of producing engraved printing rolls to produce deepened engravings:
- manual;
- Muletary;
- pantograph;
- photoelectric;
- photomechanical;
- electromechanical, etc.

In the manual engraving method, the entire pattern is engraved by the engraver directly on the copper printing rollers. The method is labor-intensive, expensive, but can be performed on special orders of the consumer or in cases where the figure can not be engraved in any other way.

The moletier method of engraving designs is also labor-intensive and involves a lot of manual work. The engraver cuts the pattern with the deepened engraving on a matrix, a small roller made of soft steel, the diameter and length of which should correspond to the pattern's rapport. If the pattern is small, the matrix is enlarged so that the whole number of rapports can fit on its circumference. The finished matrix is hardened and used to press the relief pattern onto another tempered steel roller mollet. After the mollet is hardened, it is used to extrude the deepened engraving on the copper printing roll using a mollette rolling machine. First, the engraving is rolled on the shaft in the form of a strip, the width of which corresponds to the length of the mollette, and then the mollette is moved along the axis of the shaft and rolled a second strip, and so on the entire shaft. After rolling, the engraving is cleaned by hand. The shaft is polished and chrome plated. The advantage of the mollette method of engraving is the high quality of the engraving regardless of its complexity, as well as the possibility of renewing the engraving of a worn printing roll with the same mollette.

The pantograph method of engraving is intended for drawings with a large spread. In the pantograph method, the drawing in enlarged form is transferred to a zinc sheet, which is engraved along the contours and colored according to the original. The pantograph is a special copying machine that allows the outline of the engraving to be transferred from the zinc sheet to the printing rollers at the original scale. Cleanly polished printing rolls are covered with acid-resistant mastic and mounted on a pantograph with diamond cutters, which are connected to a system of levers and manually driven when the pin is moved along the contours of the engraving. The shaft can be rotated, and thus an exact copy of the drawing is formed on its surface, reduced to normal dimensions, and the mastic layer is broken accordingly. At the subsequent processing-etching of the shaft with the solution of chlorine iron or nitric acid, in the places of the diamond-drawn pattern there is etching of copper, and a pattern is formed on it. Then the engraving is cleaned, the shaft is ground and chromed. Italian company "Olivetti" created photoelectric pantograph. It allows you to reproduce the engraving of drawings simultaneously on three printing rollers automatically, with the help of devices with photocells. They are able to "view" the photographed drawing and transmit pulses to an electronic station to amplify them. After that, they are fed to the electronic machines with diamond cutters, accurately reproducing a copy of the pattern on the surface of the shaft covered with mastic. This is followed by etching, cleaning of the engraving, grinding of the shaft and chrome plating. The entire shaft engraving process lasts about three hours.

Photomechanical method of engraving. Engraving on the printing roll is obtained by etching with chlorine iron solution or nitric acid depending on the required depth of the engraving. This method of engraving in comparison with the moletirnyi or pantorafnyi methods is more perfect and productive, is carried out with the use of photocopies of the drawing. In fact, the work of a highly qualified engraver in this case is performed by a qualified photograverkopirovshchik. This method is characterized by multistage. A light-sensitive chromogelatin layer is applied to a well-polished copper shaft, on which an exact positive copy of the drawing is carefully applied and tightly pressed and exposed to light irradiation with the help of mercury lamps for 30-60s. Under the action of light, the chromogelatin layer on the open areas of the drawing hardens, and on the closed areas does not change. After photographing the printing roll is subjected to washing with warm water, as a result of which the blown areas remain, not washed away, forming a negative image, and the unblown layer is washed away, leaving open areas of the roll, subjected to etching, after which the roll is cleaned, the engraving is corrected, polished and chromed. The method does not have a single technology, it is diverse and constantly being improved. It is characterized by reduction of the volume of manual labor, etc.

Manual, mollette and pantograph methods have lost their importance to a certain extent, but they are not completely forgotten. It is not excluded the possibility of their application in small enterprises, where the manufacture of several rollers in a month solves the problem of printing production, while the highly productive engraving technique will be doomed to downtime.

A high-performance electromechanical method is used with the MEGA machine. It consists of electronic and mechanical parts, for engraving printing rollers. The electronic part with the help of a photo-optical head provides "reading" of the original pattern pattern and converts the light flux into an electrical signal. The signal is amplified in proportion to the optical density of the "read" pattern. The received electrical impulses are transmitted to the mechanical part of the engraving device, consisting of cutting heads acting on the surface of the engraved cylinder. This method of engraving makes it possible to produce a set of three shafts per shift. For example, it would take several shifts to produce one shaft using the mollette method.

The printing elements of the roller consist of a system of ink carrier cavities. They hold the viscous ink in their cells, and the cell walls serve as a support for the rakley so that it does not deflect. The size and shape of the cells depend on fine grids of different lettering, called rasters. In terms of shape, rasters come in spine, dot, cross, and line shapes. The number of raster lines per 1 cm is approximately from 10 to 30. Rasters are numbered accordingly. The direction of the raster lines should always be at an angle to the direction of the fabric movement and to the racli touch line. The depth of the raster elements can vary from about 0.05 to 0.3 mm. By varying the depth, it is possible to change the intensity of coloration on the fabric and to obtain halftone patterns. Sometimes the drawing is outlined with a contour line, which makes it clearer. After engraving and checking, the surface of the printing roller is ground and chrome plated. This increases its durability and preservation.

The print roll press mechanisms shall ensure that the ink is fully extracted from the print. The pressure should be uniform along the roll and should be of a value equal to the minimum value that will allow the ink to be extracted. Usually, a higher value is given to the pressure, taking into account that the printing rollers must rotate the truck and transmit motion through it to the other machines in the machine. On modern printing presses, the press rolls are pressed pneumatically, hydraulically or hydropneumatically. This allows the degree of pressure to be monitored and adjusted by means of a pressure gauge. In a hydropneumatic press system, compressed air is fed into a hydropneumatic cylinder. The oil is a buffer to smoothly create the desired pressure, which is indicated by the pressure gauge for each roll on the control panel. The pressure on the print roll is typically 50-60 kN. At high values of pressure on the printing roll, one should be afraid of mechanical damage to the roll in case of a sudden and significant increase in pressure in the pressure devices at the moment of passing random thickenings (folds, knots, foreign objects) between the truck and the printing roll. In these cases, the machines are equipped with special protective devices. These allow the pressure on the roll to be relieved for a short time.

The tissue printing truck, or press, is the support for the printing rollers. It is rotated by the printing rolls through frictional forces and can be moved up or down by an individual drive, which is necessary when changing the diameters of the printing rolls and when filling the kirsa. Wide, heavy calico is often used as a lapping to cover the truck. Good lappings are cloth of linen warp and woolen weft or cloth coated on one side with a layer of rubber, as well as fabrics of synthetic fibers. Instead of lapping, some companies offer to put on the truck a pneumatic cushion, which like a tire should be inflated with compressed air. Such a cushion has permanent elastic properties.

The trafing mechanisms are used to adjust the position of the print rollers in order to ensure complete alignment on the fabric of all the design elements placed in accordance with the pattern on the individual print rollers. The need for such adjustment arises when there are two or more rollers. The pattern details will coincide if the rollers are set in parallel, without displacement of the pattern details. For this reason, trafing, i.e. adjusting the position of the printing rollers, can be done in three directions: vertically, along the axis of the roll, and around the circumference of the roll. Axial etching is usually carried out with the help of a special device capable of moving the lettering together with the shaft spike along its axis. More complicated is the etching around the circumference of the printing cylinders. Here we should pay attention to the fact that all printing rollers receive rotation from a single drive with the help of the main gear, called the master gear, but it is necessary that the multi-shaft machine installed shafts of exactly the same diameter. The shafts must receive rotation with the same circumferential speed through a gear train. If any shaft is installed in the clutch with some advance or lagging of the rapport, it is necessary, without removing the shaft from the clutch, to turn it by a very small angle forward or backward. It is desirable to carry out such adjustment in two modes: manual for rough adjustment of the idle machine and remote traflation in the process of work for final fine adjustment. Modern trafing mechanisms are characterized by the possibility of remote control.

The drive of printing presses must be able to provide a wide range of speed control. DC and AC motors can be used for printing presses, but in both cases the drive must be able to adjust speeds over a wide range, depending on the complexity and priming of the design. The printing rollers, while receiving rotation from the drive, are at the same time the driving organ of the entire printing unit, driving the truck, pickguard, cover, and fabric. In turn, on some machines, the kirsa is used as a transmission belt to transmit motion to the dryer, selective mechanism, which requires an increase in the pressure force of the rollers, leads to accelerated wear and pulling of the kirsa and adversely affects the accuracy of etching. Therefore, modern machines use a multi-motor drive system that can be automatically controlled, which eliminates the above disadvantages.

Most modern printing presses use adjustable pneumatic and hydropneumatic roll presses, advanced and automated pattern etching systems. The machines are equipped with drives for the transverse stroke of the rakley and truck lift. Modern printing machines with engraved shafts with deep engraving are characterized by an improved design of individual units and mechanisms while maintaining the basic principle of printing. The use of such machines creates favorable conditions for increasing labor and equipment productivity and improving the quality of printing, as the speed of fabric movement can be brought up to 200 m/min without fear of machine vibration.

The use of a cover made of synthetic fabrics in a number of cases even allows to refuse drying of the cover, where after washing and increased squeezing the moisture remaining on the fabric is lower than capillary retained moisture, i.e. close to swelling moisture, which does not interfere with printing. There is no need to dry the cover, which retains constant and high elasticity, its adhesion to the printing cylinder is improved, which makes it possible to reduce the degree of pressure on the rollers and even to refuse the use of burlap, replacing it with a wet-pressed cover. If necessary, the cover can be dried. Printing units of the Czechoslovak company Totex:

The unit includes an 8-shaft printing machine 12 with engraved rollers. It is equipped with remote electromagnetic etching. Nozzle horizontal dryer 2 is designed for drying fabric and cover. It has a cooling chamber 6, a cover washing unit 8 and a dryer 5. Garment washer 9 has a dryer 4.

The fabric enters the printing machine from roller 11. After printing, drying and cooling in the cooling chamber 6 it is selected from the machine by roller stacker 7. The cover also enters the dryer, and if necessary is washed on a special machine 8. It is squeezed, dried on a nozzle dryer 5. Then it is fed by the pulling pair 1 to the rotary stacker 3. Then the cover is folded into the fabric compensator 10. From it it is again fed to the printing machine. If necessary, the cover can be selected by pulling pair 1 into a cart.

The kirsa, an endless ribbon exits the printing press. It is fed to the kirzo washing machine 9. After washing, it is dried on the six-section air nozzle dryer 4. Then it is sent to the printing machine again by tension rollers, being cooled by the shop air on the way.

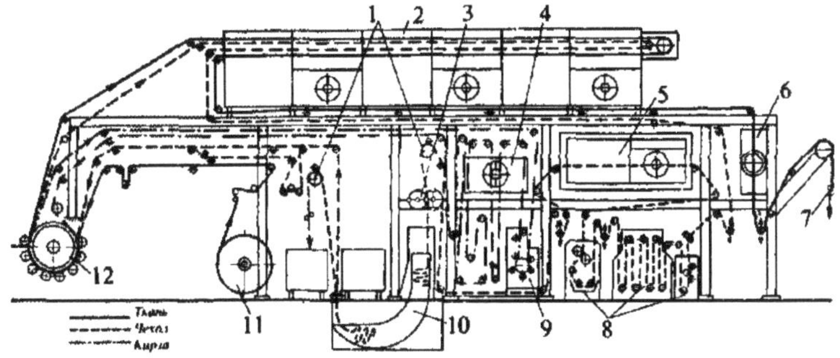

Figure 12. Tissue printing unit with engraved printing cylinders

25

The units include a kirzo washing unit, cover washing baths and powerful nozzle dryers. The kirzomoynaya drying units are used for washing, drying and cooling of the kirzha for lapless printing and are part of the fabric printing units. The printed fabric is placed on a well-tensioned kirsa, which should take on excess ink. The soiled side of the kirsa is then washed, dried, and the kirsa is again fed to the printing press, where it is tucked into the endless web. Kirza should have good susceptibility to printing ink and resistance to the action of its components, as well as easy recoil of ink and at the same time sufficient hydrophobicity, be elastic and elastic, have a sufficiently high mechanical strength and resistance to stretching, do not have a surface thickening. Now use kirza, which is a 58-layer cotton fabric or fabric made of synthetic fibers 1.5-5.0 mm thick, impregnated with special compositions that give it hydrophobic properties, strength and resistance to the action of paint components.

The printing units are available in nominal widths of 1200, 1300, 1400, 1500 and 1600 mm. A further increase in width is associated with the risk of printing roll deflection. Up to 24.5 kN of hydro-pneumatic pressure is applied to the printing roll beads, i.e. 49 kN per printing roll. The fabric speed in the machine varies from 12 to 120 m/min. Speed control is stepless from a DC drive. The installed power of the electric motors is 61.5 kW, including the printing machine 26 kW. Evaporation capacity of the dryer at a vapor pressure of 0.3 MPa is 190 kg / h, but at 0.6 MPa 240 kg / h. Overall dimensions of the printing machine, mm: 3275x3100x2350, and overall dimensions of the machine, mm: 18520x3200x5950 mm.

A number of companies produce printing presses with inclined or vertical printing rollers. Instead of a common central truck with a large diameter, they have individual small-diameter fabric trucks for each printing roll. This ensures a narrow band of contact between it and the printing roll. It also helps to increase the specific pressure in the sting of the rollers, to extract the ink efficiently from the deepened engraving, and to obtain a clearer outline when the pressure on the rollers is reduced against the normal pressure.

This arrangement of the print rollers allows the raclist to monitor all print jobs from a single workstation and control the operation of the printing unit.

The printing press with inclined arrangement of rollers is proposed by Brückner (Germany). It differs in that the printing rollers and trucks are placed close together on a fixed frame mounted at an angle of 45°. Printing rollers with rollers applying ink on a movable frame, which can be lowered parallel to the floor for changing and cleaning the rollers. The rollers are electromagnetic. The printing rollers are located in the gaps between the trucks. This allows the fabric and the kirsa to move in a wave-like motion. At the same time, the fabric web fits tightly to the surface of the rubberized pickguard, without deformation and without stretching. This makes it possible to machine easily deformable fabrics, including knitted and non-woven fabrics. At the same time, a higher trafing accuracy is obtained.

The company produces machines with the number of rolls from 3 to 10 with nominal widths from 1000 to 1600 mm. The speed of fabric movement is 5 - 80 m/min with stepless regulation. The installed power of the electric motor is 40 kW. Overall dimensions of the unit, mm: 14000x4700x4400.

The printing press of the company "Kleinevefers Jaegli" (Germany) of Saueressig system with vertical arrangement of printing rollers.

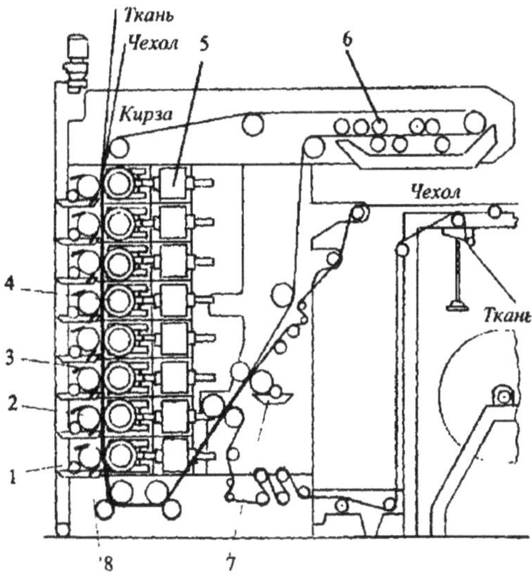

Figure 13. Printing press with vertical rollers.

Each printing roll 2 is provided with an individual press roll 4 (truck). It is located on the other side of the pickguard and has a pneumatic presser 5. The ink is applied to the printing roller by roller 1 installed in a bath with printing ink. Excess ink is removed from the surface of the printing cylinder by a rakley 8. Fibrous dyed fluff is removed by counter-rake 3. Each raclay has a pneumatic clamping device. The rake mechanism is driven by an individual electric drive. The machine uses steel, copperized printing cylinders instead of traditional solid copper jacketed cylinders. The thickness of the copper surface layer is only 1 mm.

The printing rollers are mounted on the side platens of the machine and are equipped with an electropneumatic drive with infinitely variable speed control. Side tracing is done quite accurately by hand using handwheels and graduated rulers. Vertical (circumferential) trafling is carried out from a remote control. It is possible both in operation and on a stationary machine. The unit includes a printing nozzle dryer with horizontal fabric and cover filling. The dryer is located on the same floor as the printing machine.

The fabric is fed into the printing machine from the roller and unloaded into the cart by the stacker after drying. The cover is fed into the machine from the cart and after drying is also placed in the cart. The printing machine uses the kirsa in the form of an endless web. It is cleaned from ink by means of a machine 6 for washing and drying the kirsa. The machine is provided with a device 7 for gluing a cover or cloth to the kirsa (if printing is carried out without a cover). The kirsa with fabric is pressed to the engraved rollers by means of electropneumatic devices located at the ends of the rollers.

Vertical rollers reduce the time required to prepare the machine for operation. For example, for a complete replacement of a five-color pattern, it takes 25 minutes instead of several hours. For complete etching of an average pattern it is enough to use 1-2 meters of fabric instead of several dozens, as on machines with traditional arrangement of rollers. The machines of the considered type can be used for printing transfer paper, as the web tension is minimal and does not threaten to tear the paper. The machines are available with a nominal width of 1600 mm. The maximum speed of fabric movement is 100 m/min. Speed control is smooth. The power of the printing machine motor is 28 kW, and four additional motors of 2 kW each. Overall dimensions of the machine, mm: 18500x5300x5500.

The disadvantages of the machine include rapid wear of the wetting and drying rollers, clogging of water pipe sprays.

7 Print dryers

Machines for convection drying and heat treatment of fabrics. Convection dryers are widely used in the dyeing and finishing industry for drying a wide variety of products made of natural or chemical fibers. The following types of dryers can be distinguished:
- convective-roller and curtain type with chamber-wide longitudinal blowing of the fabric;
- nozzles with V-shaped (zigzag), vertical and horizontal wiring of the fabric;
- convection-roller with local jet blowing of the fabric;
- dryers with combined fabric blowing;
- screen-drum dryers.

Convection dryers can be easily aggregated with plusovki, can be included in dyeing lines, used for drying, drying of printed fabrics, included in lines for final finishing, etc. The unification of convection dryers allows the use of a number of machines for thermal treatment, mainly by replacing steam heaters with electric ones and changing the nature of air circulation.

Convection-roller and curtain dryers with shell-blown fabrics provide soft drying, which is necessary for processing fabrics impregnated with working solutions, e.g. for nitriding, impregnation with cube dye suspension, dyeing with active or dispersed dyes, etc. Soft drying helps to reduce or eliminate migration, which improves product quality.

Air-roller machines of Russian production SVR120 with all-chamber longitudinal blowing of fabric have been widely used. The production of more advanced air-roller drying machines of MCWR brand has been mastered. On their basis a number of structurally unified MVTR machines for thermal treatment of fabrics is produced. Counterflow dryers of MCWR type (Fig. 9.6) are designed as a part of flow lines for drying or drying of cotton, linen, acetate, viscose and synthetic fabrics with surface density up to 420-500 g/m^2.

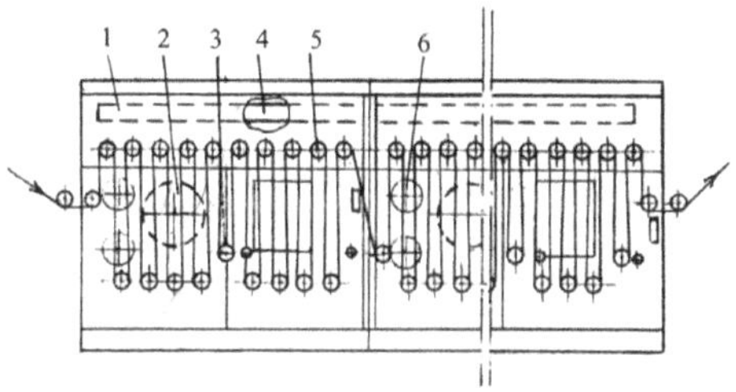

1 frame; 2 fan; 3 compensator; 4 calorifier; 5 roller; 6 - electrical equipment.

Figure 14. Air-roller drying machine type MCWD

They are available in 16 modifications with nominal widths of 1400 and 1800 mm. The modifications differ in the number of sections, the number of which is 2, 3, 5, 7 and 9, depending on the purpose of the dryer and its evaporative capacity. These machines are characterized by low drying intensity. In order to achieve sufficient capacity, the length of the fabric filling must be increased, which is about 20 m per section. The section length in the middle zone is 2080 mm. The height of the section is divided into three zones. In the middle, working zone there are rollers for transporting the fabric and nozzles for blowing it with hot air. The upper rollers have an electrodifferential drive with adjustable torque, which together with a voltage sensor ensures that the fabric is guided with minimum tension. The height distance between the rollers is 1000 mm. This eliminates the formation of bent edges and notches, which occur at a distance of more than 1500 mm, but forces to put in the middle the so-called pull rollers, which reduce the length of the free web loop. One circulation fan is installed on each side of each section. The upper and lower areas of the sections are pressure boxes with heaters for heating the air. Air circulation is carried out as follows: from the working area the air is taken by a centrifugal fan and directed to the upper and lower zones. After that, through the calorifiers it enters the nozzle devices for blowing the fabric with a speed of 20 m/s at the nozzle outlet. Depending on the modification and specific purpose of the machine, different temperature regimes are set. For example, for ten modifications the air temperature in all zones is 135 °C. In dryers of the MSRV type, the speeds are smoothly regulated by means of DC motors, the speed of fabric movement is determined by the monogram given in the handbook.

According to the method of threading, machines with zigzag, vertical, horizontal and combined fabric wiring are distinguished.

In the nozzle dryer CB6/140, the fabric filling is carried out by zigzag (V-shaped) wiring of the fabric.

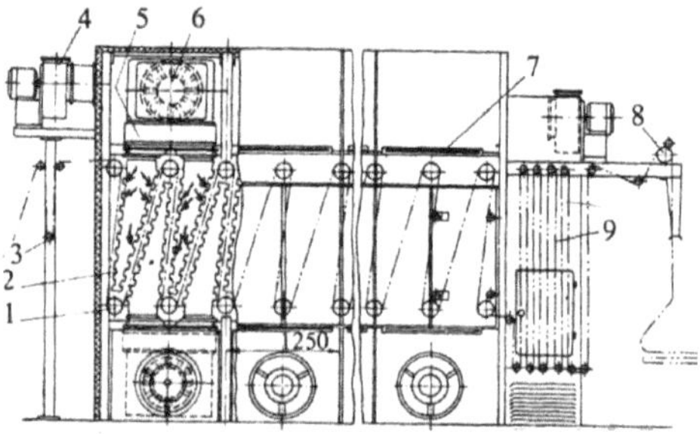

Figure 15. Nozzle dryer CB6/140 with zigzag (V-shaped) fabric wiring.

Evaporation capacity of the dryer is 540-570 kg/h. Specific steam consumption is 1.6 kg per 1 kg of evaporated moisture. The drive of the upper row of rollers is carried out by torque motors, allowing to regulate their rotation speed depending on the amount of shrinkage or stretching of the fabric. The fabric tension is kept at a minimum level. For the dryer CB 6/140 the installed power of electric motors is 94 kW. The speed of fabric movement is 16-.80 m/min. Overall dimensions, mm: 10150x3475x4050.

The disadvantages of nozzle dryers include a rather large internal volume of space occupied by boxes for nozzles, which reduces the density of filling per unit of overall length.

In convection roller dryers with local jet blowing of the fabric, the following main disadvantages are eliminated quite successfully:
- low drying intensity, which is characteristic of machines with general chamber longitudinal blowing of the fabric:
- significant loss of dryer space for nozzle installation, typical of nozzle blower machines.

In local jet dryers, the fabric is looped over two rows of rollers. Blowing nozzles are placed between the rollers.

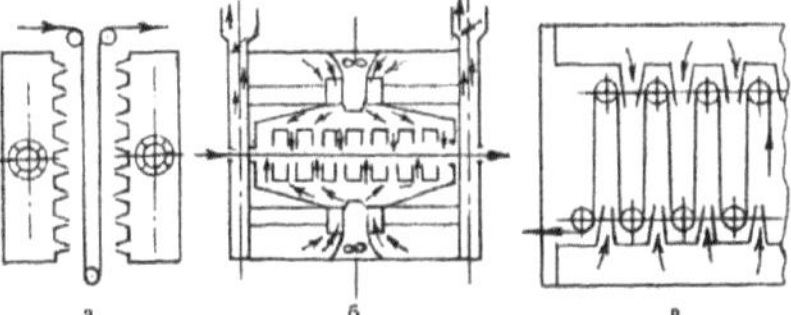

a - nozzle with vertical wire; b - nozzle with horizontal wire; c - with local jet blowing.

Figure 16. Fragments of convection dryers

By combining the methods of wiring and blowing the fabric in dryers, a high performance in terms of both evaporated moisture and fabric speed can be obtained. Since damp fabric is easily pulled out during drying with the formation of serrations, to prevent these defects, the fabric should be dried quickly at short filling lengths and then finished with long loops.

Gas drying and finishing machines are used. The disadvantages of GDF type machines include increased explosion and fire hazard, rapid contamination by combustion products of fiber dust, toxicity of gas, which requires a high level of control, and others. Probably, that is why the machines of GSO type have not been spread.

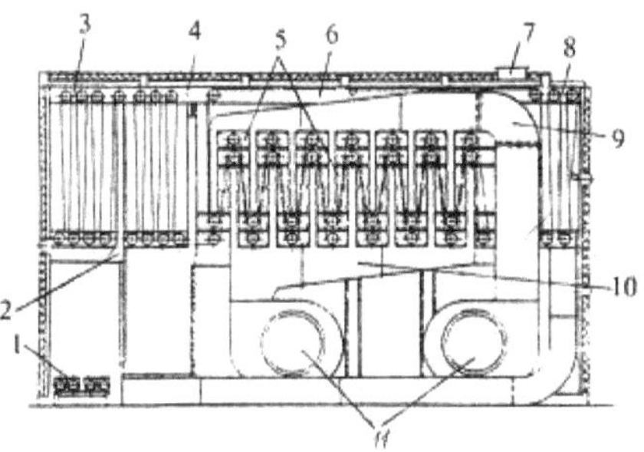

Figure 17. Gas drying and finishing machine GEO 220

In mesh-drum dryers drying is carried out by sucking hot steam-air mixture through the thickness of the textile material, so these dryers are expedient to use for drying porous textile materials with good air permeability, for example, fiber carded tape, yarn in skeins or bundles, thick but not dense fabrics, fleece, knitted fabrics, etc.

Through-suction of hot air through the fabric is the most effective aerodynamic mode of drying, increasing its intensity by dozens of times compared to the convective method. Due to their versatility, mesh-drum drying machines are produced by a number of leading foreign companies.

Mesh drum dryers are characterized by their versatility, as they are suitable for drying any textile material.

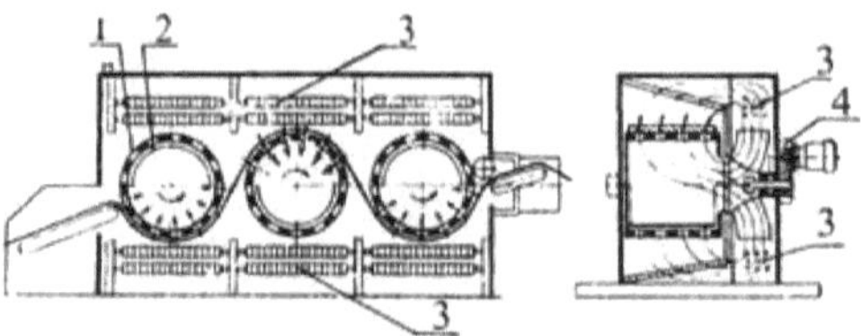

Figure 18. Schematic diagram of a screen-drum dryer with three drums.

Fleissner" (Germany) machines use powerful radial axial fans located on the end side of the drum to circulate hot air. In the machines of "Arioli" (Italy) and "Tubetex" (USA) introduced a system of intensive nozzle blowing, but it is bulky and complicates the design of the machine. Drying machines are produced with vertical and horizontal arrangement of mesh drums. They are equipped with powerful fans. They can use both single and double-sided suction of humidified air. There are mesh drum machines with a maximum girth angle using guide rollers. This prevents air from being sucked into the drum bypassing the fabric and increases the efficiency of the dryer.

Printed dryers. There are convection and contact print dryers. Their peculiarity lies in the fact that the fabric has to touch the rollers only with its underside for most of the way. Printed dryers can be installed:
- on the floor in line with the machine;
- on the 2nd floor above the fabric printing unit.

Small attic dryers are placed above the unit on the same floor.

Printing dryers of various systems contact, convection, radiation and mixed dryers are part of fabric printing units. They provide drying of the processed fabric and cover.

In convection print dryers, the fabric is loaded in a spiral pattern. At the beginning of the drying process, the fabric is tucked in with the wrong side facing the rollers. For printing dryers, the presence of cooling sections is mandatory, especially when drying inks containing exothermic reagents. It is not recommended to overheat the fabric in printing dryers. If the fabric is overheated, some ink components may decompose. The speed at which the fabric is dried determines the speed of the entire fabric printing unit. The speed of the unit can reach 90 - 100 m/min or more. Effective are printing dryers with nozzle blowing and air temperature 125 - 140 °C. Russian nozzle dryer SP1201 has vertical filling and drive from the truck of printing machine. Its evaporative capacity reaches 220 kg/h. The speed of fabric movement is up to 120 m/min. One section is allocated for drying the cover and one section for cooling the fabric. Printed dryers with a horizontal fabric run are widespread. They have an air temperature reaches 150 ° C. Produced also dryers in combination with chambers for thermal treatment with air temperature up to 220 ° C. This allows to carry out thermofixation or thermosol methods of printing. They are used for printing with active, pigmented or dispersed dyes. There is a gas dryer GSP120M. It is equipped with radiation panels and provides high evaporative capacity (250 kg/h).

Contact dryers are used much less frequently in the printing industry. The main advantage of contact dryers is their higher drying intensity. Their use is hampered by the possibility of over-drying the fabric during machine stops, which can lead to changes in the ink composition. Therefore, they are recommended for printing with dyes insensitive to over-drying. With the advent of nozzle dryers, contact drying of printed fabrics has lost much of its advantages. Air drying contributes to better preservation of printed colors. The advent of small-sized nozzle dryers with horizontal wiring allowed to place dryers above the printing and bale washing machines in the same floor. This has resulted in a significant reduction in space requirements.

8 New ways to print!!!

Textile printing for advertising purposes is gaining popularity. It finds application in a variety of formats. Any symbol or slogan of the company can be printed on the fabric. It can be used for outdoor advertising and during any marketing campaigns and events. Widespread interior items that are printed on - panels, curtains, pillows, etc. Fabric is the most popular material for making flags, and it can be both the flag of the country or republic, and the flag of the organization. The size is also not limited: there are many examples of both small decorative flags and quite large ones. With the help of printing on textiles now make even information stands. So, the original and useful from an advertising point of view element of the decor of various events becomes a textile. Fabrics for printing are used almost any depending on the purpose. It can be a delicate silk, and translucent veil, and dense satin. There are a number of characteristics that make textile printing superior to other types of printing:
- environmental safety;
- low cost compared to printing on traditional materials like paper;
- easy storage and transportation (fabric products weigh little and can be folded without fear of damage);
- long term use while maintaining the appearance;
- Washing and ironing option to keep the product clean;
- better appearance of the promotional product.

In the course of history, the craft of decorating fabric with various designs has evolved and taken on new and new forms. Nowadays, there is a huge number of ways and technologies of drawing with the help of printing on textiles. Now actively used printing on T-shirts, as well as on other clothing and accessories. Advertising posters and banners are created with its help. At the same time, in recent decades there have been significant shifts in the development of new ways of artistic and coloristic design of textile materials by direct printing.

Among the most effective and original ways of printed design should be highlighted the technology of transfer printing, which on textiles can look like a conventional printed pattern or with the use of special auxiliary materials as a velvet pattern. To date, a large number of different transfer printing technologies have been developed. In this type of textile printing, the image is transferred to the product itself indirectly. It is first applied to paper and then transferred to fabric using a heat press. Each color used in the image is transferred separately.

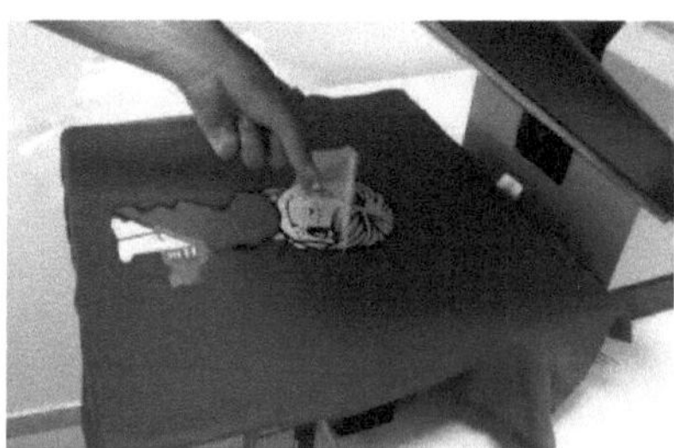

Figure 19: Transfer printing on the product

This technology finds application in the decoration of clothing items and fabric accessories. Advantages:
- the accuracy of the application of the picture;
- high quality of the resulting image;
- the possibility of creating a single instance;
- small time commitment;
- use of different materials;
- reasonable price.

As the minuses of thermal printing on textiles can be called short-lived image, which can easily be damaged during washing and ironing, as well as limitation of the color of the pattern.

Transfer printing was once a stepping stone to the creation of so-called InkJet technologies, i.e. the process of drawing directly onto fabric using large-format inkjet printers that are computer-controlled and use special liquid forms of textile dyes as inks. Now inkjet printing on fabrics is no longer a novelty, although it is still used very limitedly even in industrialized countries. This technology, along with transfer, is quite actively used only in the production of advertising products. InkJet-technologies have huge advantages in terms of image quality and speed of production of products from ideas to the finished product. Their widespread introduction is constrained, first of all, by low printing speed, which is still unacceptable for large production volumes, as well as by the high cost of digital printers and inks.

9 Foam printing

The advantages of foam compositions are that their use saves energy, water, chemical materials and reduces the negative impact on the environment. Research on the use of foam compositions for coloring textile fabrics was especially active in the 80s of the last century. At the same time in the USA, Germany and other countries the industrial use of foam technologies began. Foam coloring of carpet products turned out to be especially effective. At the same time it reduces approximately 2.5 times the consumption of heat and almost 3 times the electricity. By now foam compositions for printing with active, dispersed dyes and pigments have been developed and used. Thickener saving is 50-60%, dye yield increases by 3-12% for active dyes, by 8-13% for dispersed dyes. Significant improvement of printed fabrics' grift and increase of dyeing strength are observed.

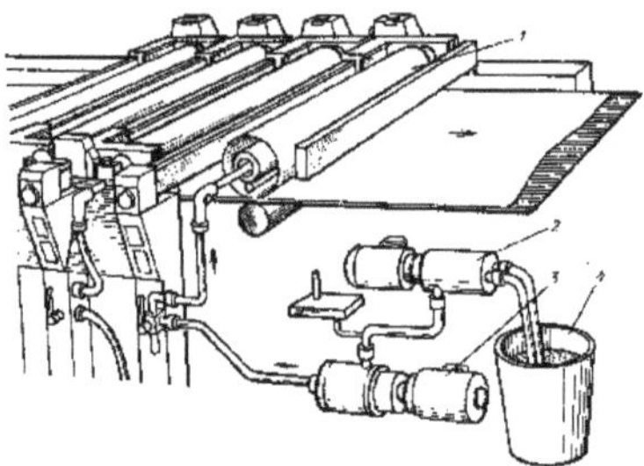

Figure 20. Diagram of the "Stork" installation for preparation of foam printing ink and its feeding into the mesh template.

Printing ink with a small content of thickener from tank 4 is fed by pump 2 into foam generator 3, where foam printing ink is formed and pumped into rotary template 1, equipped with a special rake. The experience of using foam systems in textile printing has shown that in most cases it is possible to increase the printing speed and, consequently, the productivity of printing equipment.

At present, the transition to energy- and water-saving foam technologies becomes especially urgent. Many modern brands of printing machines are equipped with special foam generating devices, which gives a real possibility of foam printing application. However, it is difficult to recommend specific compositions for foam printing of textile materials of different fiber composition with the use of certain dyes. For each variant additional research and production tests are necessary.

Original solutions and developments were obtained at the Department of Chemical Technology of Fibrous Materials of St. Petersburg State University of Technology and Design (SPUTD). The essence of foam technology of finishing consists in replacement of the most part of liquid in finishing media by air, as a result of which the moisture content of the treated material is sharply reduced and the heat and energy consumption for moisture removal in thermal treatment processes is reduced. Realization of foam finishing technology allows to significantly reduce the volume of industrial wastewater and increase the overall safety of production. For each finishing process the properties of foams should be strictly regulated and carefully controlled during the process with the help of electronic and microprocessor technology.

At foam printing the consumption of thickeners is significantly reduced: for active dyes . by 65 %; for pigments . by 50 %; for dispersed dyes by 60 %. At the same time the fabric vulture is softened, the quality of patterned coloring is improved. In case of printing with pigments the use of explosive and fire hazardous emulsion thickeners is excluded. It is possible to combine pigment printing with finishing with foamed latexes. The degree of fixation of active dyes during printing increases with the increase of foam multiplicity in accordance with an exponential dependence:

$$C_\phi = A - Ke^{-k\beta} \tag{1}$$

where
$From_\phi$. degree of covalent binding of dye to fiber, %;
A . concentration of the active form of the dye, %;
K and k . empirical coefficients;
β .- multiplicity of printed foam.

Foam printing intensifies the washing process of printed materials, reduces the temperature of their heat fixing treatment. The greatest effect is achieved in foam printing of substrate with high moisture absorption capacity (carpets, palases, pile fabrics, etc.). Textile finishing foams are produced in static and dynamic foam generators. Foams are applied to the material by special rollers, doctoring mechanisms, mesh templates and applicators of various types. Highly ecological finishing technologies are developed by leading companies in the field of textile production: FFT technology (Foam Finishing Technology) by "Gaston County" (USA), Maxi-Foam by "Kusters" (Germany), Vaccu Foam by "Monforts" (Great Britain), technologies by "Mitter" (Germany) and "Stork" (Netherlands) with the use of galvanized nickel templates and others.

The development of eco-technologies of coloring and finishing carried out at the University of Technology and Design (St. Petersburg) includes the process of patterned coloring of materials with polyvinylchloride film coatings (artificial leathers), using aqueous compositions based on textile pigments that replace expensive and scarce mineral pigments. In printing inks for this process, the concentration of organic solvents (cyclohexanone, isopropyl alcohol, alkyl acetates) is reduced from 90 to 10%. This method of printing materials improves working conditions, reduces the explosion and fire hazard of production, increases hygienic and consumer properties of products (for example, tablecloth with printed designs). Strong adhesive fixation of pigment particles on polyvinylchloride substrate is caused by physical and chemical modification of polymer coating under the action of cyclohexanone and activation of its chemical interaction with carboxyl-containing binder (film-forming) components of printing ink.

The reserve compositions used in the cold batik technique, forming the contour of the pattern on the fabric, contain benzene or white spirit as solvents for hydrophobizing components. SPUTD has developed an improved technique of cold batik with the use of water reserve compositions, which can be applied to fabrics by hand or machine (photofilm printing) methods. The production test (JSC "Sever", St. Petersburg) confirmed the possibility of obtaining high quality of artistic painting of silk fabrics while improving working conditions and reducing fire hazard in the batikization area.

One of the variants of complex processing of special purpose fabrics is the combined technology of pigment printing and final finishing when applying primed camouflage pattern. The level of indicators of special properties of fabrics achieved by using foamed pigment paints corresponds to the normative requirements of JSC "Mogotex" Mogilev. Implementation of the combined method of complex finishing of fabrics allows to reduce the number of technological operations 4.5 times, to reduce the consumption of water, steam, electricity (by 35.40%), dyes and textile auxiliary substances (by 10.15%), to exclude the use of ecologically problematic substances (tannin extract, copper sulfate, paraffin, potassium dichromate, etc.). In general, to improve the ecological situation in dyeing and finishing production of textile enterprises.

10 Printing flock on textiles

A special machine is used for printing on textiles using this technology. The fabric moves on it, while a layer of glue is applied to the place of the future pattern. Then particles of colored lint, which are called flock, are dropped on the glue. The electrified lint is fixed vertically on the layer of glue. This is how the image appears.

Figure 21. Realization of flock printing on textiles

Benefits:
- this coating is one of the most durable: its wear resistance is many times higher than the norm established by GOST. The pattern made in this technique is resistant to washing, temperatures, UV light, etc...;
- easy maintenance;
- mechanical resistance;
- impermeability. This amazing material allows you to remove liquid from the pattern before it is absorbed;
- the coating is gentle, soft and pleasant.
Disadvantages:
- flock can electrify and collect dust;
- can be damaged by exposure to alcohol;
- such textile printing is one of the most expensive.

New ways of applying designs to fabric are also emerging, such as vacuum printing. Different techniques can be used to create the same image. The choice of the textile printing method may depend on the type of fabric. Or, on the contrary, first the type of application is determined, according to which the base is selected. The choice is also influenced by the layout of the image, the number of products required, the required level of quality.

11 Textile 3D printing

Textile 3D printing is one of the modern methods of applying images to fabric. Despite the name, it has nothing to do with 3D printers and does not imply obtaining a three-dimensional pattern. Such printing on textiles creates ordinary images in 2D format. This technology has a number of advantages:
- high quality of the resulting drawings;
- application of images without any preparation of the substrate;
- durability and wear resistance of the product.

This technology is one of the youngest. It was actively used only five years ago in Russia. A unique textile printer DTX-400 was designed in Russia. It surpasses any other equipment for direct printing on textiles in all the main parameters: durability, speed, quality of the received images, etc. This printer has become actively implemented in the production of both piece products and large runs. Technical innovations associated with this printer are widely used in the creation of new models. Previously, it was believed that the quality of the picture on the finished product directly depends on the resolution of the picture. The above-mentioned printer has provoked radical changes in textile printing. With its help it is possible to print on the fabric images from 360x360 to 1440x1440 dpi. The need for higher figures was explained by the fact that the very texture of the substrate deteriorates the final quality.

The solution to this problem is equipment that prints with textile fibers. Such a printer has been pioneered at Virginia Tech by Negar Kalantar and Alirez Borhani in collaboration with DREAMS Lab. This is Flexible Textile Structures, a project that could revolutionize textile printing in the future. There is already talk that in the near future it will even be possible to make clothes using a 3D printer. Joshua Harris is working in this field. His goal is to invent a multifunctional and portable printer that most of the world's population will be able to afford.

The textile printers that are in common use today have also received a great deal of attention. One of the first was the DTX-400, followed by the DTX-600/800/900/1000 (CMYK+White). Some of them have three tables for simultaneous printing on textiles. The permissible web size is from 600x1000 mm (A1) to 1100x2100 mm (B0). Such equipment allows you to print images on fabric quite quickly and inexpensively. Features of newer models are full control and formation of a white background for drawing. If we talk about large runs of products, the time required for production can be reduced by half.

The Japanese company Brother is one of the manufacturers of textile printing machines. The GT-3 series of inkjet models allows you to print images up to 1200x1200 dpi on different types of fabrics. These printers are mostly common in small textile printing firms. Brother GT-341, Brother GT-361 and Brother GT-381 models use a special method of applying ink, which increases the speed of printing. The specificity is that white ink is applied sequentially with color ink, and the fabric goes through only one printing cycle. The big advantage of these printers is the use of eco-friendly inks that are harmless even for children. The quality of the image, applied with water inks, is not inferior to drawings obtained by other technologies. Also the durability of the final products is at an acceptable level.

A printer that allows you to print on textiles at a lower cost is the Colors SC1645TEX. It is used for business activities related to the application of images on different types of fabrics. Colors SC1645TEX is an official Russian copy of the Japanese device Mimaki JV33. The main parts of this printer are original, and therefore the quality of images is not inferior to those obtained on the prototype. The "ceiling" of resolution is

1440x1440 dpi. Changes in the design of the printer allowed to correct a number of defects existing in the original device. In particular, it is the effect of "mattress" (variation of shades), which is a constant companion of bidirectional printing. In the new device was introduced wave printing technology, which simultaneously increased both the quality and speed of image application. Thus, this printer makes textile printing of higher quality possible.

3D effect printing. Textile materials with 3D effect are holographic or three-dimensional images that are printed on the material. A good three-dimensional image is best achieved with the help of volumetric thermal applications, which are easily applied by printing. In recent years, the production of holographic photopolymer materials has begun. These films are used to apply holographic pictures to fabric. These products are available in finished form. It is produced by pressing a relief surface. The best technology for transferring 3D pictures is considered to be the method of heat pressing. This technology is simple enough to use. It takes thermal transfer film, which has a ready-made 3D pattern (3D thermal film) and with the help of a thermal press applied to the material. 3D thermal film is a combined picture, which is made by inkjet printing, as well as multicolored inserts on the basis of 3D thermal film, which imitates the volume. As a result, a 3D picture on the material is obtained.

There are two methods of creating a 3D image on fabric, these are:

1. method of application by textile printer (the image is printed immediately on the fabric of the product);

2. the method of application by means of a heat press (the pattern printed on paper is transferred to the fabric under the influence of a heat press).

The first method is effective for small productions or for individual orders. The second method is effective for large orders. The choice of fabric also plays a significant role in this type of printing. For example, the image "sits" and stays better on cotton fabric.

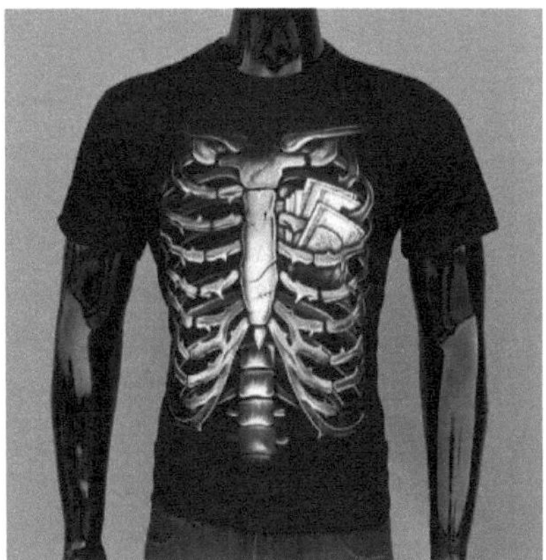

Figure 22: Textile product with 3D effect

The basic material required for holograms is a photographic plate capable of meeting the high requirement. Currently, the production of holographic photopolymer material has been developed, which plays the role of a volumetric image recording recorder. Such films are used in applying holographic images to articles of clothing. They are produced in a ready-made form. When producing them, high surface pressing should be used. Volumetric drawings can be produced with the help of 3D thermal film. This work is a combination of an image produced by inkjet printing, ink applied by an inkjet printer and multicolored 3D inserts that create the illusion of volumetricity.

The basic photographic material for recording holograms are photographic plates capable of meeting high requirements in terms of resolution. In recent years, holographic photopolymer materials have been developed and started to be produced as a recording medium for holographic image recording. It is these films that are used for application of holographic pictures on fabrics. They are available in ready-to-use form, as they are manufactured by high-precision pressing of the relief surface.

There are several ways to create a three-dimensional pattern effect on fabric:
1) using special holographic photopolymer films - the secret of volumetricity here lies in the high-precision pressing of the relief surface. The image is the most impressive and resistant to abrasion;
2) sublimation thermal transfer + 3D XPD SISER thermal film. Relatively simple. Almost any picture can be given a 3D effect.

The essence of the method: Inkjet printer is filled with special sublimation inks. The desired image is printed on special sublimation paper. By heat pressing the image is transferred to the textile material. With the help of fragments of color 3D XPD SISER thermal film "enlarged" certain details of the image and the product is again placed under the heat press. The intensity, evenness and duration of the heat press depends on the intensity of the 3D effect.

12 Transfer printing (thermal transfer printing)

Transfer printing is a type of direct printing. In transfer printing, the dye is first applied to a substrate and then transferred to the textile material to be colored by heat or other means. Application of patterned printed image using a temporary paper or film carrier of the pattern (or "from the substrate") is a modern and promising variant of technical solution of artistic and coloristic design of the appearance of various textile materials. Transfer is carried out by heating with dry heat (thermal printing) or through a water phase (wet transfer printing). The wet transfer method has now lost its importance and is practically not used in industry. Thermal printing includes sublimation and thermochemical methods.

This method in comparison with direct printing allows not only to increase the quality, but also 1.5-2 times to reduce the cost of textile materials with printed designs. This is due to the reduction of dyes and chemical materials consumption, as well as the duration of the production process, as the operations of maturing and washing after printing are excluded, which leads to significant water saving. All this contributes to solving the problem of environmental protection.

At present, there are several methods of pattern transfer from paper to textile material, developed taking into account the properties of the printed fiber material and dyes used. The most common method is based on the transfer of dye from paper to textile material due to its sublimation at high temperature ("sublistatic", etc.). Under sublimation is understood the transition of dye from the solid state to the vapor state, bypassing the liquid state. This method is used mainly for drawing on textile materials made of chemical fibers. The best results are achieved when printing on materials made of polyester, triacetate and polyamide fibers. Products made of polyacrylonitrile fibers turn yellow under the conditions of the transfer method of thermal printing. Textile materials made of chemical fibers are subjected to boiling and thermal stabilization before transfer thermal printing.

Out of many known dyes of various classes, 31 disperse dyes, 2 acid dyes, 3 cube dyes and 34 cationic dyes have sublimation ability. But of these dyes, only 39 brands are recommended for use, most of which belong to the class of disperse dyes. At present, pure shades of scarlet and bluish-green tones are not available in the range of dyes for this method of printing. Dyes used for the transfer printing method, in addition to the ability to sublimation and thermal stability in the temperature range of 180-230 ° C should have an affinity for the fiber and ensure the receipt of colors with high strength properties. The mechanism of dye transfer from paper to textile material can be represented by the following scheme:
- sublimation of dye from the paper into the vapor-air space between the paper and the textile material under the influence of high temperature and oversaturation of the vapor phase (high concentration of dye in the vapor);
condensation (transition to a solid state) of dye molecules on the outer surface of the fiber;
- diffusion of dye molecules inside the fiber;
- fixation by active centers of fiber macromolecules.

The rate of dye transfer from paper to textile material depends on the time, pressure and temperature at which the paper and textile material are in contact, the rate of diffusion of dye molecules in the vapor phase and in the fiber, the distance between the paper and the fiber material at the moment of contact, and the properties of the fiber. The pattern on paper can be applied on printing presses used in the printing industry. Special printing inks are used for drawing on paper. In addition to the dye (10-12%) and solvent (water or organic solvents - about 75%), they contain binders (synthetic resins, cellulose ethers. Transfer of the pattern from paper to textile material is carried out on two types of machines - flat presses and calendars.

Flatbed presses are batch equipment and are used for printing: semi-finished products - engineering drawings, i.e. drawings created for cut garments, coupons and piece goods.

Calenders belong to continuous equipment. Two types of calenders are used: thermal calenders operating at normal pressure and vacuum calenders, where the thermal printing process is carried out at reduced pressure. The first calender for thermal transfer printing was created in 1968 in France by the company "Sublistatik". Nowadays, equipment for thermal printing is produced by many companies in different countries. The speed of movement of textile material on thermal calenders varies within 2- 30 m/min. All types of thermal calenders have to a greater or lesser extent common disadvantages: textile products exposed to high temperatures and pressure become stiff, lose their fullness and acquire an unpleasant metallic shine.

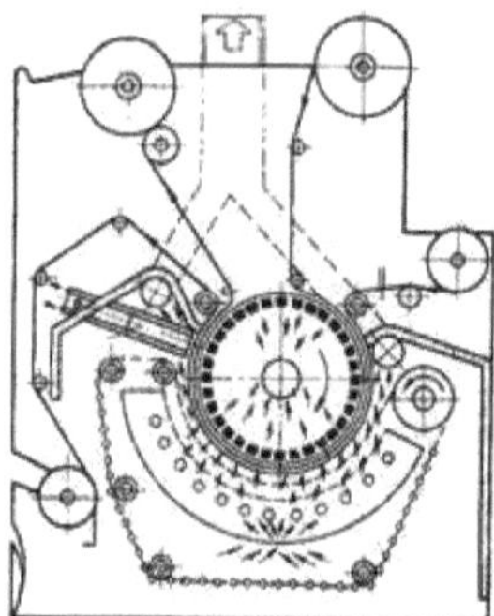

Figure 23. Schematic diagram of vacuum calender

The use of vacuum calenders allows to reduce the time required for transfer of dye from paper to textile material up to 6 - 10 s, to increase considerably (up to 90%) the transfer of dye from paper, to improve the coloring of fiber, as under the influence of rarefaction and temperature conditions are created for diffusion of dye not only to the surface of fiber, but also into its depth, which contributes to the production of bright saturated colors with high strength properties. After that, the fabric is separated from the carrier tape and rolled up into a roll, and the carrier tape is washed of dye residues, dried and the pattern is applied to it again. The ever-increasing use of thermal transfer printing does not imply a complete replacement of conventional printing methods.

Sublimation printing processes for fabrics made of cellulose fibers (or a mixture of cellulose and polyester fibers) require special pre-treatment, such as treating the substrate with synthetic resin precondensates. The dispersed dyes are sorbed by the synthetic resin, which is formed in the intermicellar spaces of the cellulose fibers during heating of the substrate with the dyes. Thus, in polyester-cellulose fabrics, the cellulose and polyester components are dyed identically. In another printing method, modified cationic dyes which have been given the ability to sublimate can be used. This makes it possible to use the "thermochromic" method for printing polyacrylonitrile fibers. Chemical plants have developed a special range of sublimating disperse dyes for printing with different transfer methods. This range includes dyes (for printing paper - substrate with engraved rolls) created for organic solvents and dyes for aqueous media (for printing with engraved rolls or cylindrical templates). Paper, plastic film or metal foil are suitable as substrates for printing. At present, mainly special grades of paper are used. Paper for the substrate should have the following parameters: surface density 70 g/m^2 ; ash content 3%; tear length 5 km; pushing resistance 30 Pa (according to Mullen); relative tensile strength 90 cN/cm (according to Elmendorf); porosity 20ml/s (according to Beck); water absorption 60 g/m^2 per minute (according to Cobb). The paper is properly prepared for printing by ensuring the desired smoothness, porosity and water absorption. This preparation is necessary to ensure contour clarity and to maximize dye transfer from the paper to the textile fabric.

In recent years, the possibility of creating a polymer film suitable as a substrate has been studied, but there is still no information about its industrial use. In one of the proposed methods, the dye is transferred to the fiber and the carrier film is volatilized. Attempts have been made to produce a temporary carrier that can be reused.

There are four methods used to print the substrate:
- flexography (high relief roller printing on paper and films);
- deep copper roller printing with oil paints;
- copper roll gravure printing with water-based inks;
- printing with cylindrical mesh templates.

Among the most modern technological solutions in the field of transfer printing a special place is occupied by *"velvet"* or **"flock printing"**. The technique of thermal transfer directly on the product ("texflock") with possible report direct printing and through an intermediate stage of production of a "picture" of long-term storage allow to receive high-quality velvet printing on textile materials of any raw material composition. Velvet printing in the transfer solution is of great interest, as it allows to diversify the design of printed materials and create velvet flock-printing effects without the use of more complex classical process of electroflocking (special lines or installations), based on traditional finishing equipment, with much better environmental protection, for example, due to the reduction of dust in the working area. This is very important for small-scale production.

Since the 1960s, the technology of velvet coating on the surface of textile and other types of materials has been intensively improved. By now, it has become an acceptable finishing method for creating functional and fashionable external printing effects, including transfer flock printing. When creating velvet printed transfer effects on textile materials over the entire surface of the fabric or in the form of stamped patterns or in small batch or piece printing, a common technological solution is used. It is based on the interaction of the textile material with a thermoplastic adhesive layer by means of which the pile is transferred to the textile from the flocked paper or film carrier under the influence of temperature and pressure. Based on the peculiarities of the application of this adhesive layer and methods of forming a velvet pattern on the textile material, the process of transfer flock printing has different technology. It can take place as follows:
- by thermo-transfer of velvet pattern from a temporary carrier to a textile material with preliminary application of adhesive composition directly on the textile, i.e. using a technically simpler solution by means of "texflock" technology, which is especially interesting for industrial realization;

- by performing an additional process step, namely, making an intermediate picture on a paper or other flock carrier for subsequent gluing (heat transfer) of such a fully formed velvet pattern on the carrier onto the textile material.

Each method has its own advantages and disadvantages. For example, the former may be of particular importance for printing across the entire width of the fabric or for the execution of rather fine or relatively thin lines of a pattern that may cause difficulties in the production of an intermediate picture for heat transfer. The application of the adhesive layer on textiles is practically no different from the usual process of printing with ink. An important technological advantage of this method is that it allows you to perform multicolor pictures due to the possibility of using velvet printing in rapport with conventional pigment printing, both on light and dark backgrounds. The level of strength performance with this method of flock printing corresponds to the category "conventional". In case of application of technologies of creation of an intermediate picture on a temporary flock carrier, the decorative pattern can be performed only as a single-pass printing, as the picture bears a local adhesive coating, which ensures its adhesion to the textile at the place of its overlay. Here rapport printing, especially with industrial methods, is practically impossible. But it is possible, for example, to apply a velvet pattern on both smooth and printed background of the material. A significant advantage of this technology, especially if the transfer flock pictures are made with the use of special fusible thermal powders applied in the process of manufacturing these pictures, is a higher resistance of the resulting velvet patterns to washing at elevated temperatures, as well as to dry cleaning. It is possible to obtain a multicolor pattern, but only by creating this pattern as a stencil pattern with white flock. But such a multicolor picture will actually also be intermediate for the subsequent transfer of its entirety to the textile material. In this case, a two-stage printing process is carried out. First, on the flock coating on a temporary film carrier is performed almost conventional color pigment printing in accordance with the coloristic solution of the picture, and then the entire resulting dry color print is applied to the common substrate of the adhesive layer and fusible granulate. Color printing is carried out with specially selected pigment compositions, providing deep penetration of the ink inside the pile coating, in order to avoid "unprinting". The process of adhesive coating in the manufacture of intermediate heat transfer picture also requires special attention, as it performs a number of functions. The adhesive coating should include to a certain length the particles of lint, which are, thanks to the temporary adhesive layer, on the carrier, securely fix them and the layer of fusible powder and, thus, provide the velvet picture with durability during its thermal transfer to the textile. The operation of thermal transfer of velvet flock onto textiles is usually recommended to be performed on thermal presses of various types in the temperature range of 170 - 190 °C for 10 - 30 s at medium values of pressure on the textile material. Then, for all variants of flock printing, the carrier is separated from the material, but only after the material has cooled down completely at the textile-flock contact point. At the present stage of development of transfer printing technologies, fundamentally different methods of forming such pictures for applying them on textile materials have been developed and are also used. These methods allow to completely exclude from the process the stage of printing by traditional methods with the use of templates. This applies to various transfer technologies, including velvet patterns.

13 Computer technology in printing

The main components of computer technology serving the needs of printed textile materials production are:
1) Computer-aided design systems for creating textile designs;
2) software for creating technological programs for controlling digital printers;
3) software to perform color calibration of digital printers;
4) software for creating technological programs for controlling digital engravers;
5) Digital engravers for making templates;
6) Paint formulation software;
7) color measuring equipment;
8) paint mixing equipment;
9) Digital printers for layout;
10) auxiliary textile equipment for modeling and small series (small-sized rippers and heat presses);
11) Standard related computer equipment (scanners, graphics tablets, networking equipment, storage).

The use of computer technology allows to solve the problem of reducing the time from the idea to the finished sample as a result:
- It becomes possible to respond instantly to fashion trends and increase the variety of fabrics produced;
- The cost of creating each new design is reduced by reducing the number of artists on staff.

14 Digital fabric printing

In contrast to rotary and flat screen printing machines, digital screen printing machines use dyes in the form of ready-to-use liquid inks. The specifics of using textile inks is that the fabric for subsequent digital printing must be prepared in a certain way. The fabric must be dried and rolled. The application of TVB is usually done with a platen or stencil drum with large interstitials. For this reason, the process of preparing the fabric for digital printing does not require any additional equipment, as it fits into the standard screen printing process.

Digital inkjet printing machines print on dry fabric. Good absorption of liquid ink by the prepared fabric and moderate heat drying of the prints ensure that there is no smudging. The digital inkjet printing machine is filled with ink of all base colors at once. Usually six to eight. Printing inks are mixed directly on the surface of the fabric, creating the desired shade on each section of the image. The printing process is controlled by a RIP-station - a computer with a special printing program - RasterImageProcessor (RIP - RasterImageProcessor). The digitally printed fabric can be fixed in the core mill of the main production. The process of fixation of active and acid dyes on protein molecules of fabric after digital printing is almost similar to the process of their fixation after classical screen printing. The difference is that the dye fixation occurs to a more complete extent. Liquid dyes in the printing process are applied more sparingly and are well absorbed by the fabric fibers. The main production equipment can also be used to wash the finished fabric. Special compact equipment for impregnation, fixation and washing of the finished fabric can be used to organize a closed digital printing production.

Pigment inks for printing on fabric contain a binder in addition to water and dye. Therefore, printing with pigment dyes does not require the pre-application of any RTV to the fabric. However, in order for the dyes to be better fixed on the fabric fibers, hydrophobic application to the fabric must be avoided before printing.

For high-quality digital printing with water-based disperse inks, a thickener (e.g. synthetic starch) must be applied to the fabric beforehand. Without this, the ink will spread on the smooth polyester fibers, the borders of the plates will become uneven, and color transitions will be blurred. After printing, the drawing should be fixed in a tunnel oven or width-drying machine at a temperature of 150-160 degrees Celsius for 3 5 minutes. In drum drying or calandrovoe thermopress at a temperature of 180-200 degrees for 30-45 seconds. In this case, dispersed dyes penetrate into the polyester fibers to a depth of several micrometers, forming a so-called solid solution. Only the pre-applied thickener remains on the fabric surface after fixation. Since the amount of dry thickener on the fabric impregnated for digital printing is only 1-2 grams per square meter, even the final rinsing of the fabric is not required in all cases.

The advantages of the technology are simplicity, high brightness and durability of prints. The disadvantage is that it cannot be used for drawing on natural fabrics.

Modern digital fabric printing systems do not simply print finished images onto fabric. They make it possible to reproduce future fabric samples with high accuracy, even without making templates for classic flat or rotary printing. As input information, a set of electronic files of rapport separations is used, including not only flat colors and speckles, but even images. In addition, information about the colors selected by the colorist for each of the separations, absorption coefficients for each of the colors and the sequence of their superimposition is taken into account. The layout of the design by the digital printing system is done automatically taking into account the rapport scheme. The correct formation of true colors is ensured by the color combination module. The resulting prints reproduce as closely as possible on full-scale samples the design developed by the designer and color schemes prepared by the colorist. If necessary, both the design and its color schemes can be changed and the samples can be printed again. Only after final approval of the samples can templates be produced and mass production can begin.

15 Inkjet printing on fabrics

The history of technology of inkjet printing on fabrics began in 1997-98, when the Swiss company Ciba, introduced to the market inks for printing on cotton and linen fabrics. The development was tested on Encad's NovaJetPROe plotters. By 1998, similar ideas had captured the minds of developers from various technical fields. The search went in several directions:
- Adaptation of aqueous inkjet technologies;
- application of inkjet technologies on non-aqueous inks;
- development of thermal transfer printing technologies;
- Adapting traditional textile technologies to inkjet printing.

The history of inkjet plotters began with the use of aqueous inks designed for printing on papers and films with special coatings. The most beautiful solution was to find the possibility of printing with the same inks on fabrics. This approach best suited the interests of ink and plotter manufacturers, as they did not need to create new products. The focus was on the use of special impregnations, fixing the image, providing its resistance to mechanical influences and water. At the same time any finishing treatment should be excluded. Such fabrics with appropriate impregnation already exist, but they have not yet become widespread. The problem is not only in their high cost, but also in the capriciousness of the technology. The ink manufacturer can slightly change the formula in the next batch, and the fixation properties of the impregnation will change.

Plotters on solvent inks have become in recent years a kind of standard solution for the production of outdoor advertising. Interesting plotters have appeared, working on UV-bonded inks. Attempts to adapt them for printing on fabrics are understandable. But it is not widely spread due to various difficulties:
- fabrics sealed with such inks lose their softness and naturalness properties (a "crust" is obtained);
- image fixation is not strong enough;
- hygiene requirements are not met.
- There may be problems with pulling the roll fabric through the plotter due to the structure of the fabric.

The essence of the technology of image creation with the help of thermal transfer is simple: the printing is made on a conventional plotter, the same conventional ink on special paper, then in a thermal press or even a laminator the picture is transferred to the fabric. The whole trick is in the "thermal transfer" multilayer paper itself:
- a layer of paper backing;
- polymer layer;
- inkjet layer.

In the thermal transfer process, the image layer together with the polymer layer is transferred to the fabric and the paper backing is separated. The polymer layer seems to laminate the image, forming a protective coating. This technology is often confused with sublimation thermal transfer printing. Meanwhile, these are fundamentally different technologies. On sublimation will be discussed below in other sections. With all the simplicity of the method, it is used quite rarely. Among its disadvantages:
- high cost of specialty paper;
- "crust" on the fabric (the fabric loses its softness);
- insufficient ink fixation strength.

Today, several types of specialty textile inks have been developed that are used to print on the following types of materials:

- active - for cotton, linen and viscose fabrics;
- acidic - for silk and wool fabrics;
- pigmented - for cotton, linen and viscose fabrics;
- sublimation (disperse) - for artificial (polyester) fabrics.

Active and acid inks give excellent quality, but require impregnation of the fabric and, after printing in the plotter, steam fixing and rinsing. Until recently, the problem was the lack of equipment for low-volume production. In 2001, Rimslow (Australia) produced a "ripper" for steam fixing. In 2002 - equipment for impregnation of fabrics. It is not cheap, but still its price is many times less than industrial equipment for textile factories.

Pigment inks are interesting because they do not require pre-impregnation of the fabric, but are fixed by heat treatment in a heat press or "zrelynik". They are technologically advanced. The problem is limited color coverage compared to active, acid and sublimation inks.

There are two ways to use sublimation ink: direct printing (fabric requires impregnation) and printing on thermal transfer paper. The latter option is the most common. Almost all textile technologies require impregnation. Everywhere - heat treatment. Somewhere - washing with water after fixing. It turns out that there is no simple solution.

According to Digital Textile, a magazine covering digital computer technology, approximately 300 million meters^2 of fabrics are printed by inkjet each year, in more than 400 facilities, each with 6 inkjet printing presses. Inkjet technology is based on 3 main elements: printers, programs, and ink/dyes. As for the programs, they are now built for full color space using 6 - 8 dyes. Ciba started commercial production of inks for all types of substrates and their mixtures back in 2004. The following brands are now available for consumers:

Cibasron MI - for cotton, linen, viscose based on active dyes,
Terasil TI - for transfer and direct printing on polyester fabrics,
Terasil DI - for direct printing on polyester fabrics,
Lanaset MI - for polyamide, wool, silk based on acid dyes,
Irgaphor TBI - for cotton and blends according to the pigment printing principle.

In order for printers to operate smoothly, the ink must have certain properties and keep these properties constant for one or two years of possible storage. The list of requirements for this type of ink is as follows:
- colorants and pigments must have a small particle size;
- chemical and physical parameters of the composition must be stable;
- printing ink must have the necessary viscosity, surface tension, must pass easily through filters, must have a stable pH value.

Inks provide durability comparable to that of conventional printing, as they are made on the basis of existing dyes, inheriting their properties, and often have the same chromophores as conventional dyes. Manufacturers of printing equipment have mastered the production of industrial printers, the printing speed of which does not yet exceed 150 m/h. This indicator is the main deterrent to the widespread introduction of digital inkjet printing.

Proponents of computer technology believe that it is not correct to compare a single InkJet printer and a traditional printing press only in terms of speed. It is more correct to consider the total result. The print quality in terms of color and resolution is an order of magnitude higher than that obtained with traditional pattern printing.

InkJet printing is cheaper than conventional printing on small batches. A rotary press costs at least ten times as much as a single InkJet machine.

Figure 24. Print quality by color and resolution

One person can operate several printers connected to one computer. A traditional printing machine today is operated by several people, and in addition, we need to take into account the staff of workers in the ink cooker and template making area. All this should speed up the introduction of modern printing methods.

Figure 25. Modern printing methods for production

The implementation of InkJet technology in textile manufacturing can look as follows:
- the first stage - introduction of digital technologies at the stage of production preparation and organization of work with the customer according to the principle: "Today sample - tomorrow production";
- The second stage is the introduction of InkJet technologies into mainstream production.
Direct digital printing on textiles is one of the most modern types of printing, which is characterized by a combination of efficiency and economy. This technology uses special printers. Image transfer is similar to conventional printing on paper or other standard media, but requires a fundamentally different printing unit, as well as special water-based textile inks colored with pigment dyes and containing special polymers. These polymers give the ink the ability to hold securely on the fabric. When printing on light-colored fabric, the image is applied by a printing press. Then it is pre-dried, covered with a special primer (to increase the dye's resistance to physical and chemical influences). After that, the final drying of the product is carried out, during which the polymerization of ink and primer takes place. When printing on dark fabric to the above process is added one more stage - first of all on the fabric special white paint is applied backing for the future color image, exactly matching the contour with the main picture.

52

The layer with the substrate undergoes a thorough drying and then the same algorithm is used to print light-colored fabrics. Printing on dark fabrics is a more labor-intensive process with the use of additional consumables, requiring more time for the production of each product. The complete process of processing and printing one product on one printing machine takes about 10 minutes, in the case of dark fabrics - 20-30 minutes.

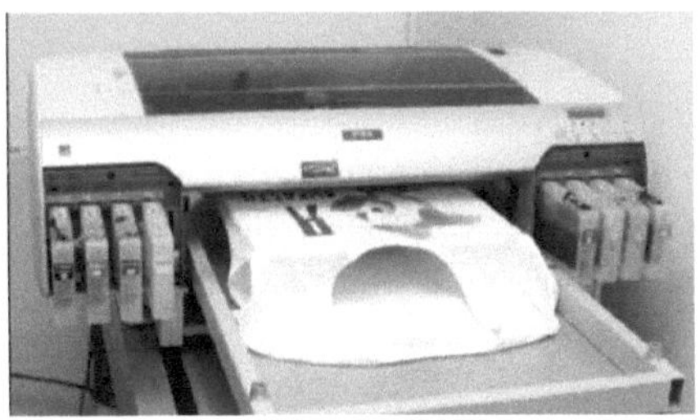

Figure 26. Direct digital textile printer

This equipment is filled with water-based ink and atomizes it onto the fabric without any intermediate step. The image is then fixed with a heat press. This type of printing is especially effective for natural fabrics, but can also be used for synthetic fabrics. In order to obtain a dark-based pattern in this way, it is necessary to first apply a light-colored ink underneath it.
Benefits:
- small time commitment;
- full-fledged print application;
- environmental safety of materials;
- the ability to create drawings of any size;
- high image quality combined with a relatively low price and no minimum or maximum limit on the number of print runs;
- optical print resolution from 1440 dpi (actual resolution depends on the texture of the fabric);
- the level of detail is much higher than thermal transfer and silk-screen printing;
- The image can withstand more than 60 washes without significant loss of quality, as polymerized dyes become literally part of the fabric;
- special effects are available, such as gloss and smooth gradients, which cannot be printed with silkscreen, for example;
- The image does not feel to the touch - the fabric is freely breathable, the T-shirt is comfortable for daily use;
- the image will not crack on stretch fabric, as the thread itself is colored and the printed area remains as stretchy as the fabric itself.
Digital equipment is understood as devices that print directly from electronic files received from workstations, and uses neofset technology, and direct ink application technology, as in printers and risographs. It would be more correct to call this method of printing "printing without the use of real printing plates". Image information is transferred from digital media. In offset printing, the image is transferred using offset plates. Digital

printing allows you to personalize each print without the additional cost of equipment changeover.

Figure 27. Direct paint application technologies

The advantages of digital printing are cost savings, and wide pattern possibilities.

Digital printing is generally defined as any printing process that uses computerized electronic files to print a product consisting of raster dots, toner or ink. Many of the manual steps that are inherent in traditional printing processes can be avoided with digital technology.

Digital printing technology can be classified into two categories: variable printing and direct printing.

Wide format digital printing. Wide format printing is used for the production of outdoor and interior advertising.

Figure 28. Outdoor advertising production

The printing width of such machines can reach 5 meters, and the length - tens of meters, the machines use the principle of inkjet printing. The material used for printing is paper, banner fabric, mesh, special textile materials. The range of equipment manufacturers is very wide.

16 Digital sublimation printing

Sublimation printing - printing in which the ink at a temperature of 180-200 °C and pressure is transferred from the carrier to the surface to be painted. In direct sublimation printing no intermediate medium is used. Sublimation or sublimation - the transition of a substance from a solid to a gaseous state without staying in the liquid state. Printing is applied to products made of synthetic white fabric (on colored, the color of the fabric is mixed with the color of the print). This method of printing allows to apply photographic images with high quality to the fabric. The technology allows you to get bright colors that are resistant to environmental influences. This technology is used for printing on sports uniforms, printing patterns for ribbon embroidery on gabardine and satin, production of flags, flags, pennants and other souvenir products.

The print becomes resistant to environmental influences only if the object of transfer is a synthetic fabric. That is, if you transfer the image sublimation method on a mixed fabric and just after washing it, the more faded the image the greater the content of cotton in the cloth. Printing is done as follows: the image is printed in mirror image on sublimation paper with special sublimation ink. Printing can be done both on a household printer, filled with such inks, and on special large-format plotters. Then the product on which you want to print, is placed in a heat press, on top of it is placed printed on paper image, printed side to the product. After that, under the necessary pressure and temperature the paper with the print is pressed by the thermopress to the product for a few seconds. The image is transferred from the paper to the fabric. The temperature and sublimation time depend on the type of ink, paper and the product itself. Printed on paper print is used once, as the ink is almost completely transferred to synthetic fibers. This method is widespread in the printing of advertising and specialty t-shirts. For this purpose, special T-shirts made of 100% polyester or two-layer T-shirts (inner layer - cotton, outer - polyester) are used. However, walking in such clothes is uncomfortable and it fulfills more the role of souvenir or advertising (worn by promoters for a short time, over other clothes).

Digital sublimation wide format printing on fabric begins with sublimation, or disperse, inks being filled into a wide format inkjet printer. As a rule, they have an aqueous or oil base. Manoukian, InkTec and a number of other companies produce such inks.

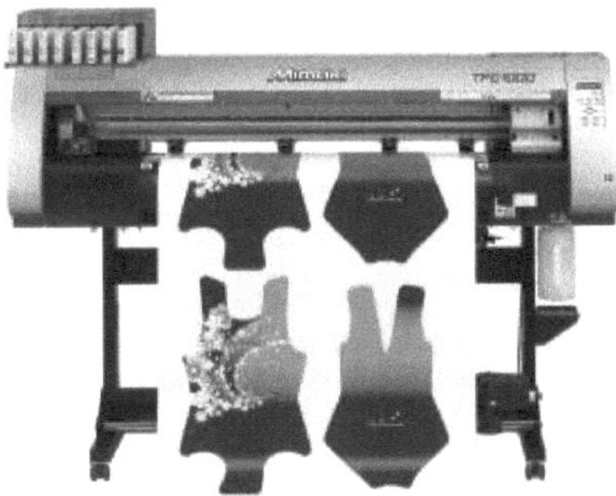

Figure 29. General view of the digital textile plotter

Sublimation dyes allow large format printing on fabric, on various polyester surfaces, including synthetic or blended polyester fabrics (polyester), such as lavsan. The image, applied using sublimation ink, is resistant to mechanical and temperature effects, light, washing. Therefore, sublimation large-format printing on fabric is the most organic technology of printing on textiles.

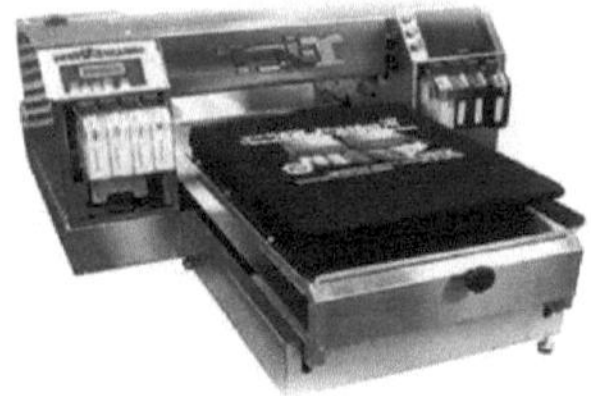

Figure 30. Use of sublimation dyes

It is no coincidence that long before the advent of sublimation inks for large-format inkjet printers and, consequently, the use of this technology in the advertising industry, it began to be used in the industrial production of fabric, <u>clothing and textile products.</u>

Digital large format printing on fabric means a whole new approach to the production of textiles in general.

Figure 31. A new approach to textile production

In recent years, a new paradigm - "firstsell then produce" - has been gaining momentum in all segments due to the development of digital technologies. This is evidenced by the spread of 'digital' in various consumer markets, e.g. in photographic printing, book publishing. Textile markets, including <u>clothing</u> and <u>interior</u> design, are next in line.

Sublimation printing on fabric is traditionally a process that uses an intermediate medium - sublimation paper. It is on it that the printer prints, after which, with the help of special equipment, the ink is transferred to the fabric or textile product.

Figure 32. Outputting a drawing to sublimation paper

The technological process of sublimation printing on fabric consists of several stages:

1) Printing on sublimation paper

First, the file, pre-flipped mirrored, is printed on sublimation paper. The main feature of this material is that it does not absorb ink, allowing the next technological stage to transfer them almost completely to the fabric. The most well-proven brands of sublimation papers include Transtjet Texprint.

Caution. Do not judge the quality of color reproduction by the image printed on sublimation paper. It will be significantly inferior in brightness to the one you will get on the fabric at the end of the technological cycle.

2) Transferring a printed image to a fabric or textile product.

Further, the print obtained on sublimation paper is combined with the front side of the fabric or product on which the image is to be applied. Actually for the transfer, depending on the shape and size of this product is ***used flat or calender heat press.*** This device under the influence of temperature and pressure transfers the image to the fabric, as a result of which the dye is embedded in the structure of fibers and securely fixed in them.

Benefits of sublimation:

- pattern quality;
- Durability and resistance to environmental influences, including washing and ironing;
- the possibility of creating a single instance;
- variety of materials used.

Disadvantages:

- is only available for printing on fabric containing synthetic fibers;
- the base can only be white in color;
- high cost.

To realize sublimation digital printing on fabric, you need the following equipment and software:

(a) A wide-format inkjet printer that prints with dye-sublimation ink, such as the Mimaki JV33 or JV34;

b) a flat or calender heat press, such as the SuperTex Model SR;

c) professional RIP (PhotoPrin versions 4, 5, 6, RasterLink) and a program for building ICC profiles.

Theoretical explanation of thermosublimation. During thermosublimation under the influence of high temperature there is a local sublimation (ablation) of ink from the intermediate medium and its deposition on the printing material.

As an intermediate medium, porous thermosublimation paper is used, the surface of which is subjected to a special chemical treatment. This paper must absorb a large amount of ink and then evaporate it with maximum efficiency.

Sublimation is the transfer of a substance from the solid phase to the gaseous phase without an intermediate liquid phase. The process is controlled by varying the temperature and/or the duration of the thermal energy pulse: depending on the amount of thermal energy applied to the image element, different amounts of colorant are transferred to the print material. In doing so, the diameter of the image element remains approximately the same, while the optical density varies. The transfer of the gaseous dye to the print material occurs by diffusion (Dye Diffusion Thermal Transfer).

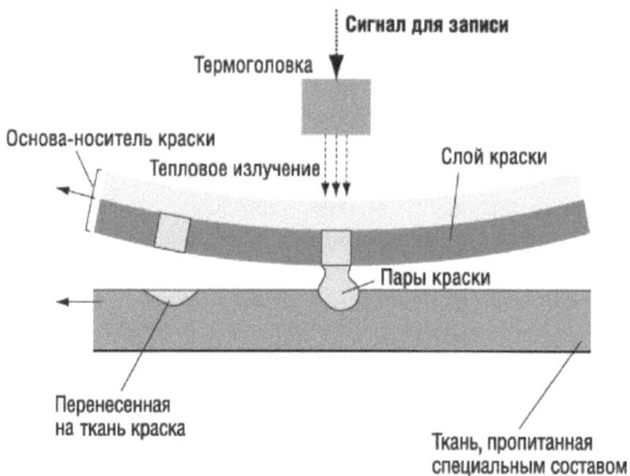

Figure 33. Dye transfer to fabric in digital printing

Currently, the indirect printing method is used to apply images to polyester (polyester) fabrics, to composite fabrics containing at least 50% polyester or acrylic fibers, as well as to materials with a special coating. A film or spray can be used to create such a coating.

When heated to about 165° C, the polyester molecules seem to open up and the sublimation ink, which is in a gaseous state, can diffuse into them. When the product cools down, the polyester molecules slam shut, locking the cured ink in place. The result can be compared to a tattoo on human skin.

Sublimation dispersion inks are highly viscous and are able to vaporize and bond with polyester at high temperature. However, these properties of sublimation inks limit the range of plotters capable of high-quality and stable operation with them.

Sublimation printing: advantages. If we talk about the advantages of sublimation printing, among them are:
- printing without washing;
- environmental friendliness of the technology;
- high quality images;
- fade resistance;
- rich and deep colors;
- colors can easily withstand multiple washings;

- Ability to print complex and multicolor images quickly;
- the ability to print any print run;

Sublimation printing: disadvantages. Of course, the method of sublimation printing has not only advantages, then some disadvantages. To the disadvantages of sublimation printing can be attributed:

- the need to comply with certain climatic conditions during the technological process;
- high image quality can only be achieved with a light-colored woven base;
- printing is only possible on fabrics with more than 50% synthetic fibers.

Indirect digital printing on cotton fabric. Indirect printing method includes two technological steps:

1) applying a mirror image to an intermediate medium (sometimes called thermal transfer);

2) transferring an image onto fabric under the influence of high temperature.

The first step can be performed using digital printing devices such as inkjet printers. The image is transferred using thermal transfer methods.

Thermal transfer paper accepts ink when printed and then gives it away in a heat press to the item to be decorated. Thermal transfer of image from paper to fabric takes place with the help of heat presses at temperatures from 180 to $240^0 \circ$ C. The temperature depends on the thickness and structure of the fabric, thickness of the thermosublimation paper, as well as on the physical and chemical features and colorimetric characteristics of the ink. Industrial thermal presses fall into two categories: flatbed (flatbed) and rotary (calender); the latter work with roll material.

Thermal transfer from thermal transfer paper allows for the reproduction of tonal multicolor images on fabrics. This can also be realized with thermal transfer paper and an iron.

The following should be considered when conducting the work of using digital printing:

A) The *file requirements for printing* (sublimation) on fabric are not much different from the standard requirements for large-format printing on paper. However, due to the fact that fabric is a fiber weave, very fine details (fine lines, patterns, small print) may not look as sharp as on paper. The texture of the fabric can also improve the quality of the image. The customer's source files do not always correspond to the required resolution, and those errors that would be striking when printed on paper will be smoothed out on the fabric just due to the textural qualities of the material.

Raster files: formats PNG, JPG, GIF, PSD; color mode RGB, CMYK; size 2340x3310 px (for A3 size) resolution 200 dpi. vector files: formats EPS 8 version; model CMYK; objects without stroke; text is translated into curves; not allowed to fill type "mesh" and gradients (replacement by bitmap).

Figure 34. Paper-based drawing for reproduction on fabric

B) The perception of the image is also ***greatly influenced by the*** properties of the selected fabric. That is, the same digitized image (file) printed on different <u>materials</u> will be the same in terms of color rendering, but will look different. So, when printing on a shiny fabric, such as satin, the color is so saturated that it becomes even brighter than the laid pontoon. And if the picture is printed on a matte material, then the saturation of the colors is brightened. Color rendering can also be influenced by such a characteristic of the fabric, such as whiteness. White can be a warm shade or a cool shade, as well as be dazzling or calm. Taking all of this into account, in conjunction with the other fabric characteristics (the key ones being density and mechanical properties of the fabric), different fabrics are selected for different applications.

B) Lint and dust can nullify the efforts of technologists, and color always changes shade when exposed to temperature.

D) Transfer is printed with special sublimation inks, pressed tightly and heated to the temperature at which the transition of the dye to the fibers of the fabric. The mechanism of coloring here is such that the paint does not lie on the surface of the product, and penetrates between the fibers and settles on them a thin layer. As a result, the properties of the fabric itself do not change, and the print is super resistant to washing.

E) Some materials (for example, the same acetate silk) except for sublimation printing can not be printed at all, because they are not susceptible to traditional dyes.

Transfer of the image to the fabric under the influence of high temperature. The second stage can be performed using special electro-mechanical devices, so-called heat presses.

Figure 35. Heat press for transferring a picture from sublimation paper to fabric

Description of thermopress and its working principle. Thermopress is a special device that allows you to apply from paper to non-standard media: mugs, T-shirts, plates, souvenirs, puzzles, magnets, ceramics, metals, glass.

With the help of a heat press (under the influence of temperature), the image from the paper is transferred to the desired surface.

The heat press is a technologically complex product, therefore it is necessary to observe all operating conditions, otherwise it will lead to the failure of the heat press.

A set of print nozzles:
-for fabrics and T-shirts:
- for the mugs;

- for round emblems and plates.

Specifications:

Voltage: 220 V

Heat table 29x38cm

Maximum temperature: 225° C

Time range: 0 ~ 999sec

Overall dimensions 588*380*450mm

Procedure for working with the heat press:

1. Make sure the cord is well connected to the wall outlet. Place the fabric swatch on the press. Place the image paper with the image facing the fabric. Adjust the desired pressure and turn on the power.

2. Set the desired temperature and time. The temperature will start to rise.

3. When the temperature rises to the set temperature, the buzzer emits a signal.

4. The time counter is then switched on. As soon as the set time has elapsed, the buzzer sounds again. The top of the heat table must be raised. This stops the buzzer signal. The work is finished

For a better understanding of how the heat press works and how to use its attachments, we recommend watching this video tutorial: ECH-800 8 in 1 Combo Heat Press Machine for T-shirt, Plates (2 sizes), Mugs(4 sizes), Caps - YouTube [360p].

Questions for control

1. Give a characterization of digital printing.

2. How does file quality affect print results?

3. Can digital printing be used for double-sided printing?

4. What advantages do you think digital printing offers over traditional grid pattern printing?

5. Give an explanation for the high resistance of coloration of textile material in digital printing to various physical and chemical influences?

6. What is a transfer and how to use it?

7. What are the conditions of working with a thermo press when realizing digital printing?

17 Materials and colors for printing

Most often synthetic fabrics are used for modern textile printing. There are many types of them, each with its own weight, texture and material ratio. They are a good basis for transfer printing due to their high durability. Also, these fabrics can be used for printing 2x3 banners and other outdoor advertising products in large and small sizes. Photo printing on textiles can also be carried out using natural fabric as a base. Of course, many people prefer to wear clothes made of such materials for environmental and practical reasons. But, if we talk about drawing, then for this natural fabric needs to be prepared in a special way. As for the methods of such printing on textiles, it can be thermal transfer printing at a sufficiently high temperature, direct printing and screen silk-screen printing. These are the methods that are used to print designs on cotton.

The choice of fabric depends on what is being made and what your budget is. If we are talking about clothing, then, of course, you should give preference to more natural materials. If it is some stretching or banners, it is more reasonable to use synthetic fabrics, as they are more resistant to the influence of precipitation, mechanical damage and other harmful effects. The price of printing on textiles will be formed depending on the chosen method of image application and the size of the product.

The types of paints that are used for painting on fabric are plastisol paints and water-based paints (water-based paints). Both the former and the latter have their advantages and disadvantages. If we talk about with which ink printing on textiles will give the most environmentally friendly and hygienic drawing, it is certainly more natural water-based inks. However, the technological features of imprinting on fabrics lead to the use of plastisol ones. They differ from water-based paints in that they do not dry out under the influence of room temperature. This makes it easier to work with grids of different numbers without blocking the cells when printing the smallest details of the pattern. However, the production of inks does not stand still. Now on the market there are both plastisol inks, which are not inferior to water-based in terms of safety of composition, and water-based inks, which show greater stability when used for printing patterns on textiles in various ways.

Plastisol and water-based inks are most commonly used for printing on natural and synthetic canvas varieties. However, solvent-based inks, also called solvent inks, can also be used. Textile for solvent printing is an artificial material, which, in addition, is treated with a water-repellent composition. It can also be a blended fabric. Solvent inks are often used to depict symbols on workwear. Printing on textiles is carried out not only by direct drawing, but also with the help of such a method as transfer printing. As mentioned above, this method implies the presence of an intermediate layer, which is usually a special paper. With such printing on textiles there is a possibility of using all three main types of colors. Although some advantage still remains for plastisol.

Pigments. Pigments are used in the printing of textile materials from all kinds of fibers and their mixtures, because the fixation of pigments on the fabric is not due to intermolecular and chemical bonds, but by means of a binder. This fact is one of the main advantages of pigment printing. Pigments are inorganic (oxides of various metals, metal powders, etc.) and organic coloring substances insoluble in water, acid and alkali solutions. Inorganic substances such as zinc oxide, titanium dioxide, ultramarine, soot, metal powders, etc. can be used as pigments. Organic pigments include mainly two groups of compounds: pigments proper, i.e. coloring substances that do not contain groups that give solubility, and varnishes. Among the pigments, monoazo- and disazo-pigments form the major part, and some polycyclic cube dyes, some thioindigoids,

phthalocyanine pigments, nitroso-pigments, luminescent pigments, etc. are also used. The spectrum of organic pigments consists of: water insoluble azo-, anthroquinone, indigoid, phthalocyanine and other dyes. Currently, many textile companies choose the most economical and environmentally friendly flushless pigment printing methods for the widest range of textile materials made of cotton, hydrated cellulose fibers, as well as blends of these fibers with polyester fibers. Advantages of pigment printing:
- availability of a wide range of colors with high lightfastness and lower dye consumption ($\geq$ 15%);
- simplicity in technical execution;
- reduction of all types of discharges into the environment;
- Significant reduction of water consumption and energy costs due to the elimination of the washing operation;
- the possibility of obtaining various coloristic effects;
- suitability for all types of printing equipment: from flatbed photo film printing to rotary printing with high printing speeds;
- efficiency of application in foam technology using systems of foam generators and RSH
- rakel of "Stork" company.

Pigment dyes have no affinity for the fiber. Absolutely insoluble in water. They do not penetrate into the internal structure of the fiber. In aqueous dye baths or as part of printing inks they form highly dispersed systems. They are fixed by adhering to the outer surface of elementary fibers with the help of a special binder, which forms a strong dyed film at the stage of thermofixation. The main requirements for the quality of pigments are a high degree of dispersibility and homogeneity. Pigments should contain 95-97 % of particles 0,5-2,0 microns in size. Otherwise, their volume will exceed the thickness of the formed film of the binder, which will lead to a decrease in the friction resistance of the color. The degree of dispersibility determines not only the purity and saturation of the coloring, but also its resistance to friction. Pigments for printing are in the form of aqueous highly dispersed pastes containing 25-40% of colorant, surfactants and antifreeze.

The quality of pigment-printed textile materials depends on the printing compounds used. Their main components are pigment, binder (film-forming component), fixative (mesh-forming component), softener, emulsifier, thickener. Different companies offer their own compositions for pigment printing. The technology of pigment printing of all textile materials is as follows:
- application of the printing composition on machines with grid templates (flat or rotary
- depending on the type of printed material);
- drying in the print dryers of the respective printing presses;
- heat treatment in dry kilns in superheated steam environment or on heat treatment lines with dry hot air at 140-160°C for 4-2 min.
Pigment printing materials include: pigments, binders (film forming component), fixatives (mesh forming component), defoamers (defoamer), emulsifiers (dispersant), softeners and thickeners.

Binder. The binder, which is a film-forming compound, is of great importance for fixation of pigments on the surface of textile materials. The main requirements to the binder: under conditions of heat treatment it should form a film on the surface of elementary fibers. The film must have the following properties:
- must have a high adhesion capacity, i.e. the ability to retain pigments in the film structure and on the fiber;

- must be transparent, elastic, non-toxic, resistant to mechanical influences, to light and to dry cleaning.

Currently, three groups of binders are mainly used: acrylate, butadiene and polyurethane. Each of them has its own advantages and disadvantages. In practice, compositions based on thermoplastic and thermosetting polymers are used as binders. Thermoplastic polymers are derivatives of acrylic acid. They serve for the formation of a transparent film. They are called *film-forming polymers. Thermosetting* polymers - methyl derivatives of urea and triazine - are used as "crosslinking" reagents. They provide a strong bond between the film and the fiber. Dye particles are distributed inside the film. They are called *mesh-forming reagents.*

Acrylate-based binders are the most commonly used binders in textile factories. This type of binder has good light fastness, resistance to high temperatures, washing and dry cleaning, but low resistance to dry and wet friction. To improve these characteristics in pigment printing, it is recommended to use acrylic binders mixed with butadiene binders. It should be taken into account that butadiene polymers are very sensitive to temperature and light. As a result of these effects, yellowing of the polymer film occurs. Therefore, mixed binders containing butadiene binder should only be used for dark shades and should not be used for clear, light and pastel shades. Binders based on polyurethane provide high print quality even at relatively low fixation temperature (110-120°C), but have a rather high cost.

The composition of the printing ink includes:
- fine pigment;
- complex binder (thermoplastic and thermosetting polymers);
- a catalyst of an acidic nature;
- thickening;
- various kinds of textile auxiliary substances (TAS) that improve the quality of films, colors and consumer properties of finished fabrics.

After printing and drying the fabric is subjected to heat treatment at temperatures 140-160° C for 3-5 minutes depending on the nature of the fiber. Rinsing after printing with pigments is not required, as the degree of their fixation is close to 100%. This significantly simplifies the technological cycle of printing. However, a problem arises when selecting a thickener. It must be removed from the fabric during drying and heat treatment, otherwise the fabric acquires additional stiffness in addition to that provided by the binder film.

Thickeners. Emulsion thickeners, which are viscous solutions of organic solvents not miscible with water, can be used as thickeners. Its components, due to volatility, are completely removed from the fabric during drying. However, there is a risk of contamination of wastewater and the atmosphere. At present, ammonium salts of polyacrylic acid and their copolymers are mainly used as thickeners. The use of ammonium salts is preferable because during drying and thermofixation after printing, the ammonia volatilizes and the polyacrylic acid itself remains on the fabric.

The advantages of using such thickeners are:
- Polyacrylic acids are characterized by poor solubility in water and do not swell, which has a positive effect on the resistance of printed designs to physical and chemical influences;
- they serve as catalysts for polycondensation of crosslinking agents (fixatives) and their interaction with film-forming agents (binders).   Disadvantages of using synthetic thickeners:
- They are very pH-dependent and sensitive to electrolytes.

For this reason, the amount of synthetic thickener used also depends on the quality of the production water, the colorants used, and the electrolyte content of the excipients used.

Fixatives (crosslinking agents). They are added to printing pastes to improve resistance to friction and washing. This is especially important when printing on fabrics made of synthetic fibers and their blends with natural fibers. They are derivatives of urea- and melamine-formaldehyde resins with the lowest possible free formaldehyde content. An acidic environment is required for the interaction of binders and crosslinking agents. In most cases, this environment is created by the use of synthetic thickeners, if ammonium salts of polycarbonic (polyacrylic) acids are used.

Softeners. Because of the binder, crosslinking agent and thickener, the fingerboard inevitably becomes stiff. Softeners are used to modify the veneer of the fabric in the desired direction. Softeners have an optimum softening effect while slightly affecting the print strength values. Fatty acid esters, mineral and silicone oils are used as softeners.

Defoamers. They are used to prevent foaming during the preparation of printing pastes or during the printing process itself.

Emulsifiers (dispersants). In addition to their emulsifying action, emulsifiers are stabilizers of the printed pigment system. They prevent agglomeration of pigment particles, allow redispersing dried binder particles, thus improving the flowability and permeability of printing pastes.

Full pigment compositions are used for printing textile materials made of synthetic fibers and their blends with natural fibers, as well as for fabrics with particularly high quality requirements, e.g. for printing military camouflage fabrics. When printing linen fabrics, textile companies minimize the pigment composition and use mainly three components: binder, thickener and dispersant.

Table 1
Printing Ink Recipes

Company name	Composition	Concentration, g/kg
Huntsman (Ciba)	Water	Up to 1,000
	Pigment	X
	Lyoprint PB-HC (48%) (binder)	60-150
	Lyoprint PT-XN (thickener)	12-14
	Lyoprint PFL (retainer)	0-15
	Lyoprint PA-NS (defoamer)	1-3
Tanatex	Water	Up to 1,000
	Pigment	X
	Nofome SR 500 (defoamer)	0-2
	Tanasperse EM (emulsifier)	2-4
	Tanabond EP 2015 (binder)	60-160
	Tanaprint EP 2078 (thickener)	20-28
Minerva	Water	Up to 1,000
	Pigment	X
	Antifoam W conc (defoamer)	1-3
	NH4OH (25%)	1-2
	Binder	80-180

	Clear GST (thickener)	14-17
	Finish S (emulsifier-plasticizer)	10-30
	Fixator (C/N, LF, NFO) (fixator)	5-15
	PrintoFin SUPRA (rheology modifier)	5-10
Clariant	Water (soft)	Up to 1,000
	Printofix pigment	X
	Printofix Binder 83 (binder)	50-180
	Printofix Fixierer WB (fixer)	0-15
	Printofix Softener HP (softener)	0-15
	Printofix Thickener CSN (thickener)	10-16
BASF	Water	Up to 1,000
	Pigment	X
	Helizarin Binder ET-95 (binder)	70-80
	Luprintol MCL (emulsifier)	10-20
	Lutexal HIT (thickener)	23-25
	Luprimol SE (softener)	0-10

Printing ink is prepared by simple mixing of the formulation parts. It may also contain emulsifiers, softeners, defoamers, etc.

Approximate recipe for the preparation of printing ink (in g/kg):

-pigment (paste)   60-100;

-methazine   100;

-ammonium chloride with water (1:3)   25;

-25% ammonia solution   10

-thickening;

After each component has been added, the printing ink is mixed thoroughly. If the pigment is taken in powder form, it is pre-mixed with water at a ratio of 1:1. In this recipe as binders used metazine and available in the emulsion thickener latex SKS-65-GP, catalyst - ammonium chloride, stabilizer - aqueous solution of ammonia. Dibutyl phthalate can be added as a plasticizer, silicone can serve as a defoamer, etc. Printing inks from pigment dyes are recommended to prepare relatively liquid. Thick can dry on the template or shaft of the printing machine. Therefore, it is recommended to add ethylene glycol (up to 30 g/kg) to the printing ink. For better impregnation of hydrophobic fabrics, introduce wetting agents.

The fabric printed with pigments after drying with the utmost precaution ensuring good ventilation of the dryer and eliminating the danger of ignition of white spirit vapors, which are in emulsion thickener in significant quantities, is subjected to heat treatment. At this time polymerization of film-forming resins and fixation of pigments on the fabric takes place. The method of heat treatment depends on the properties of the film-forming resins used. It can be limited to simple steaming for 20-30 minutes at a temperature of 100-105 ° C or steaming in thermal chambers at 120-140 ° C. Sometimes drying of fabric after printing is combined with thermofixation, passing it through hot drying drums. It is not recommended to wash the fabric immediately after fixation. It is necessary to keep it for 24 hours. In general, fabrics printed with single pigments do not require washing. This is one of the advantages of this printing method. However, it should be noted that thorough washing with detergents increases for some pigments the resistance of colors to friction. Pigment dyes are mainly used in direct printing - they give bright, even, pure tones with a clear outline, high resistance to light and wet handling, but not always with sufficient resistance to friction and wet wiping. Pigment dyes can be used in rapport with dyes of different classes, but the mandatory washing of fabrics in this case reduces the coloring effect. When using white pigments on fabrics get beautiful matte patterns, and when mixing white pigments with colored pigments - matte colored patterns. For the preparation of printing inks with white pigments, there are a large number of widely varying recipes. For example, one of them is for fabrics made of synthetic fibers (in g/kg):

-50% polyvinyl acetate emulsion 400;
-10% polyvinyl alcohol thickener 200;
-MF-17 resin or metazine 100;
- TiO_2 paste with glycerin 1:1 150;
-dibutyl phthalate 120;
-ammonium hydrogenate 30.

The printed fabric is thermofixed on a frame with infrared radiation at a temperature of 150-170 °C at a rate of 7-10 min. Then it is sent to finishing.

Printing with metal powders is a frequent application. Approximate recipe (in g/kg):

-50% polyvinyl acetate emulsion 6500;
-10% polyvinyl alcohol thickener 200;
-MF-17 resin or metazine 100;
-bronze powder 100;
-dibutyl phthalate 80;
-ammonium hydrogenate 20.

The fabric is fixed in the same way as for white pigment printing.

Active dyes are preferably used for fabrics made of cotton, viscose and high-modulus hydrated cellulose fibers for children's use, as well as for articles subjected to frequent washing and abrasion. In addition, active dyes are indispensable for filling fabrics made of natural silk and wool. The main method of printing fabrics with active dyes is direct printing, which can be carried out by one- and two-stage methods.

One-stage printing technology includes the following operations: printing, drying, steaming or "dry" heating of fabric, washing with water, treatment in hot detergent solution, again washing with water and drying. Fixation of the dye in the fiber after printing is carried out by saturated water or superheated steam, dry hot air, IR-rays. Depending on the reactivity of active dyes and fixation temperature (100 - 200°C) the duration of heat treatment varies from 30 s to 5 - 10 min. Printing ink contains dye, sodium bicarbonate, urea, ludigol, thickener and water. Sodium alginate as well as starch and cellulose esters are used as thickening agents. Hydroxyl-containing compounds close in chemical structure to cellulose, such as starch, should not be used as thickeners. Active dyes may interact with hydroxyl-containing thickeners, resulting in undesirable printing effects such as reduced dye retention or the formation of a water-insoluble film on the printed areas of the fabric that makes the fabric stiff to the touch. Alginate thickeners also have their disadvantages. The first disadvantage is that they do not provide high dye color yield. This leads to an increase in the concentration of dye in the printing ink. The second disadvantage is the very short shelf life of the prepared thickener, especially in summer time. Natural alginate thickeners are susceptible to bacteria, which for 2 - 3 days lead the prepared thickener in a non-functional state. Synthetic thickeners are devoid of these disadvantages.

Urea is added to printing inks to increase the solubility of active dyes. It also acts as a medium for fixing the dye to the fiber. This function is realized at the stage of heat treatment, when urea, melting, plasticizes the thickener and thus creates conditions for the transfer of the dye from the printing ink to the fiber. The effect of urea depends largely on the nature of the dye and its concentration in the printing ink.

Ludigol is a weak oxidizing agent and is introduced into the composition of the printing ink in order to prevent destructive effects on the active dyes in the medium of the matrix and the cellulose fiber itself.

The two-stage technology of printing with active dyes provides at the first stage application of dyes containing no alkaline reagent to fabric and drying, at the second stage - impregnation of fabric with alkaline electrolyte solution, heat treatment and washing, similar to washing in one-stage technology. Thermal treatment of fabrics for dye fixation is carried out the same way as in the one-stage method, but the fabric enters the heat chamber in a wet wrung state (80% residual humidity). Making the treatment with alkaline solution in a separate operation allows to use stronger alkaline reagents and, consequently, to accelerate the fixation of dyes at the stage of heat treatment. Unfortunately, the two-stage technology has not been widely used due to the lack of wet rippers at textile enterprises.

Monochlorotriazine dyes, less frequently vinylsulfone dyes and extremely rarely bifunctional dyes are used in active printing. There are quite a large number of different types of active dyes with different structures of active groupings in terms of chemical structure and properties. The most common types of active groupings include cyanuric chloride, vinylsulfone and pyrimidine derivatives. Nowadays, there are also bi- and polyfunctional active dyes that contain two or more active groupings. The reactivity of the active groupings increases in the following series: monohaloydtriazine, di- and trihaloydpyrimidine, vinylsulfone, dihaloydtriazine and bifunctional.

Table 2

Types of active colorants from leading companies

Active grouping	Manufacturer (country)	Name of dyes
Vinylsulfone $Kp-SO_2$ H_2C-CH_2 OSO_3H	NPO KATION (Russia)	Active T
	Dystar (Germany)	Remazol
	Kisco (South Korea)	Synozol
	Sumitomo (Japan)	Sumifix, Inozin V
	Everlight (Taiwan)	Everzol
Difluoropyrimidine/vinylsulfone (structure)	Dystar (Germany)	Levafix CA
Chlorotriazine/vinylsulfone	Kisco (South Korea)	Synozol K Synozol HF

		Synozol SHF
	Sumitomo (Japan)	Sumifix Supra Sumifix Supra EXF Inozin FS
	Clariant (Switzerland)	Drimaren CL Drimaren K
	Hunstman (Ciba) (Switzerland - USA)	Cibacron W
Fluorotriazine/vinylsulfone	Hunstman (Ciba) (Switzerland - U.S.)	Cibacron C, Cibacron FN
Difluorochloropyrimidine	Clariant (Switzerland)	Drimaren R Drimaren X Drimaren HF
Monochlortriazine	NPO KATION (Russia)	Active
	Hunstman (Ciba) (Switzerland - USA)	Cibacron P
Dichlorotriazine	NPO KATION (Russia)	Active X
Monofluorotriazine	Hunstman (Ciba) (Switzerland - USA)	Cibacron F

(chemical structure: triazine ring with Kp, N, R, N, N, F substituents)		
Fluorotriazine/fluorotriazine (chemical structure: triazine ring with Kp, HN, N, N, NH, МОСТИК, F substituents)	Hunstman (Ciba) (Switzerland - USA)	Cibacron LS
Bifunctional (monofluorotriazine/monofluorotriazine, vinyl sulfone/vinyl sulfone, monochlorotriazine/vinyl sulfone)	Hunstman (Ciba) (Switzerland - USA)	Cibacron H
	Everlight (Taiwan)	Everzol ED
	Dystar (Germany)	Remazol Ultra RGB
Polyfunctional (molecule patented)	Hunstman (Ciba) (Switzerland - USA)	Cibacron S

The possibility of obtaining optimal results when printing with active dyes is determined by the complex of their properties:
- solubility;
- reactivity;
- by the degree to which they're held in place by the fiber;
- by the affinity of the dye for the fiber.

These properties must be taken into account when selecting dyes for printing processes. High solubility allows the dye to be dissolved in a small volume of water, thus avoiding the liquefaction of printing inks.

The reactivity of the dyes shall be sufficient to ensure dye fixation in a technologically acceptable time with a sufficiently low to moderate amount of alkaline agent. The reactivity should be relatively low to ensure stability of the printing dyes during storage and to preclude dye fixation outside the printed pattern when the printed fabrics are washed. The affinity of active printing dyes should not be high. The higher the affinity, the more difficult it is to wash off unfixed dye. This increases the risk of printing faults such as background staining.

Monohaloydtriazine and vinylsulfone are used for printing. Bifunctional ones are rarely used.

Table 3
Some active colorants recommended for printing

Manufacturing company	Name of dyes
Clariant	**Drimaren P** Yellow P-5GL Yellow P-3RL Orange P-R Red P-BN Red P-2B Purple P-2RL Blue P-3RLN Blue P-RL Turquoise P-CN Navy blue P-2RLN Black P-N Black P-BLN
NPO KATION	**Active** Bright yellow 5Z Orange 5K Bright red 6C 120% Purple 4K 120% Bright blue K 120% Blue 5K Green 2J Olive 2J Black K Black 4ST
KRATA Tambov NPO	**Active** Yellow 2KT Yellow translucent 2KT Golden yellow KT Bright orange 4KT Orange JT Scarlet 4JT Red 2ST Red-purple 2KT Bright purple 4KT Bordeaux 4ST Bordeaux ST Red-brown 2KT Black 4ST

Printing with active dyes. Printing of cellulose fiber fabrics *with active dyes* can be carried out by one-stage and two-stage technology.

In *single stage printing, the* composition of the printing ink includes:
-colorant;
- sodium bicarbonate;
- urea;
- thickener;

- water.

Active colorants must not interact with thickeners. The thickener must not contain free -OH groups. Used: sodium alginate, CMC. When choosing a thickener, the following should be taken into account: starch thickening agents should not be used, as they contain hydroxyl groups that can react with the active dye. As a result, the resistance of dyes to wet processing and friction is reduced and the fabrics become unpleasantly stiff to the touch. Therefore, only substances inert to the dye should be used as thickeners. Sodium alginate is best suited for this purpose. It contains carboxyl groups that prevent reaction with the active dye. Cellulose esters, such as carboxymethylcellulose - CMC, can also be used. Urea is added to increase the solubility of the dye. In addition, urea plasticizes the thickener, promotes fiber swelling and increases the moisture content of the fabric. Additionally, urea acts as a medium in which the dye reacts with the fiber. After application of the printed pattern, the fabric is dried, treated in a ripper to fix the dye on the fiber, washed and dried. Dye fixation can be carried out with saturated water steam (steam dryer), superheated steam and dry hot air (thermal dryer) or IR radiation (radiation-thermal chamber).

Two-stage printing *technology* provides:
- application of an alkaline agent-free printing ink to the fabric and drying;
- impregnation of the material with an alkaline electrolyte solution followed by heat treatment.

Heat treatment is carried out in a high-speed ripper for 20-40 seconds at a temperature of 120-125° C, where the wet-pressed fabric is exposed. Then it is washed to completely remove hydrolyzed dye and thickener from the material. Washing is carried out with cold water, detergent solution at boiling, hot and cold water.

Active dyes used in printing have special requirements:
- easy washability of the hydrolyzed dye without staining the white background of the fabric;
- high coloring and covering power;
- high durability of the resulting colors.

Most active dyes containing the azo group (-N=N-) are capable of etching. Therefore, they are used in etching printing. They can also be used in reserve printing for coloring the main background. They are most often used for color reserves under black aniline or on the background of insoluble oxyazo dyes. Printing with active dyes can be combined with printing with dyes of other classes.

Features of washing of fabrics printed with active dyes. Printing ink for printing cellulosic materials with active dyes usually contains dye, sodium bicarbonate, urea, thickener (sodium alginate, carboxymethyl cellulose, emulsion thickener) and water. After heat treatment, the active dyes are on the fabric in the form of chemically bound to the fiber and in the form of hydrolyzed and adsorbed dyes that have not reacted with the fiber. The hydrolyzed dye on the fabric does not have a bright color and slightly changes the shade. Therefore, washing is the last final technological operation of processing the printed fabric before its finishing. It has a decisive influence on the quality of the finished fabric and resistance of its colors to wet processing and friction.

The purpose of washing fabrics printed with active dyes is to remove unfixed hydrolyzed and adsorbed active dyes from the fabric, to wash out thickeners and auxiliary substances that did not react when the dyes were fixed on the fiber.

Active colorants can interact with hydroxyl-containing thickeners. This may cause undesirable effects during printing:

- decreases the degree of dye fixation;
- A water-insoluble film of thickener is formed, which makes the fabric hard to the touch.

Thickeners should not form insoluble compounds with dyes and other constituents of printing inks and should be easily removed from the fabric after fixation of dyes during washing. To remove hydrolyzed dye and thickening, washing is carried out at a temperature of 50-60 ° C in a solution of surfactant. When washing printed fabrics, the substances on the surface of the fabric are most easily removed. In the case of active dyes characterized by good solubility, the main factor limiting the rate of dye removal should be considered the presence of a thickener. If the thickener is slowly and completely removed from the fiber, the dye must diffuse into the washing solution through the thickener layers, which sharply slows down the mass transfer process. The ability of the thickener to interact with the dye has a significant effect on the quality of the washing. This leads to a decrease in the degree of dye fixation. An increase in the amount of dye unbound to the fiber significantly complicates washing. Therefore, for better removal of unfixed dye, thickeners and dyes that do not interact with each other should be used, or conditions should be selected to reduce their interaction. To ensure good washing quality, the following conditions should be observed:
- increase the degree of dye fixation;
- remove obstacles to the diffusion of the dye from the fiber material into the solution. Increased temperature and the use of special detergents have a positive effect on the rate of diffusion and dissolution of dyes.

Printing of fabrics with cube dyes. Wide range of colors, brightness and high resistance of colors to light, wet processing and friction make cube dyes one of the most promising in printing of fabrics from cellulose fibers. Printing is carried out according to three fundamentally different schemes:

1) printing ink containing pre-reduced dye, followed by drying and color development in a steamed burner (alkaline-hydrosulfite-rongalite *method*);

2) printing ink, containing unrecovered cube dye in a finely dispersed form, reducing agent and potash, with subsequent drying and manifestation of coloring in a steam burner (*rongalite-potash method*);

3) printing ink containing unrecovered cube dye in a finely dispersed form, followed by drying, treatment with an alkaline solution of reducing agent and color development in a high-speed ripper (*two-phase method* - analog of the suspension dyeing method).

In all three methods the process ends with the fabric passing through washing machines, where oxidation, soaping and washing are carried out. The most promising and technologically favorable of the mentioned printing methods is the rongalite-potash one. At its realization the printing ink contains cube dye in the form of paste, rongalite, potash with water and thickener. After printing and drying, the fabric is treated in a steamed reducing burner at 103-105° C for 10 minutes. This is followed by washing. This method is not suitable for printing with ordinary cube dye powders. In this case the hydrosulfite-rongalite method is used. Cubic dyes are used not only for direct printing, but also for etching and reserve printing. Reducers are used as etching and reserve substances. The composition of printing ink and the technology of patterned coloring in this case is similar to direct printing on white fabrics. The amount of rongalite is increased in this case, because it is spent not only to convert the cube dye into a soluble form, but also to decolorize the main background in etching printing and to reserve coloring in the reserve method.

Printing of fabrics with cubozols. Cubozols are widely used in printing of cotton, viscose staple and blended cellulose-containing fabrics. They make it possible to obtain bright colors resistant to all kinds of physical and chemical influences. Along with the nitrite method of color manifestation, the *steaming method* is used in printing. It contributes to obtaining bright colors without processing fabrics in sulfuric acid. The use of sulfuric acid causes corrosion of equipment. When realizing the steaming method in the printing ink introduced sodium chlorate ($NaClO_3$), ammonium rhodanide (NH_4 CNS). For better dissolution of the dye, solutium salt is added. To improve the stability of printing inks, ammonia is introduced, as well as water and thickener. After applying the printed pattern, the fabric is dried and treated in a steaming oxidizing burner for 5-7 minutes at a temperature of 102-$103°$ C. Then it is washed, washed, washed again and dried. Cubesols are used in reserve printing with cubesols and active dyes for coloring the main background of the textile fabric.

Printing of fabrics with azoic dyes. Two printing methods are used in industry:
- with thickened diazor solutions on pre-nitrogenated tissue;
- by a mixture of azotols and stable passive forms of diazo compounds.

The first method is used in ground printing. Printing ink is applied to the fabric impregnated with azotolate solution and dried. The ink contains: diazole, sodium acetate or acetic acid, thickener and water. After printing and drying, the fabric is sometimes treated in a zrelnik to accelerate the reaction of azo-combination and to obtain brighter and stronger colors. Next, the fabric is washed. This method of printing has a number of disadvantages:
- Azotol is wasted inefficiently and is difficult to wash off the fabric;
- it is difficult to obtain a variety of colors, as only the nitrogen component can be varied;
- additional tissue nitrification surgery is required.

These disadvantages led to the creation of the second printing method, when azo- and diazo compounds are applied to the fabric as part of one printing ink. In this case, the diazo compound is in an inactive form, which is not able to react with azotolates. Combination occurs only after the printing ink is applied to the fabric, by converting the passive form of the diazo compound into the active form, which combines with azotolates. Several variants of realization of this method have been developed, of which the method of printing with neutrally manifested *diazaminols* (pologenes, neutrogens, diazaminols with mark H, etc.) is the most widely used. Diazoamino compounds in these preparations are stable under normal conditions, but under conditions of steam baking at high humidity and temperature, they decompose with the release of active diazo compound. It reacts with azotolate to form insoluble azo pigment on the fabric. Printing ink contains: pologen, urea, sodium hydroxide, magnesium sulfate and thickener. The fabric is printed, dried, steamed and washed. On fabrics dyed with azoic dyes, white and colored printing patterns are obtained by etching and reserve methods.

To obtain white etching, a printing compound containing rhongalite, anthraquinone in the form of a paste with glycerin, potash ($K_2 CO_3$), thickener and water is applied to the fabric. The printed fabric is dried, passed through a steam ripper and washed.

Cubic dyes are used *to produce colored etchings*, whereby they are converted to a soluble form in the grain mill and oxidized to an insoluble pigment at the washing stage.

To obtain white patterns by the reserve method on the background of azoic dyes, reducing agents or sulfuric acid salts of zinc and aluminum are introduced into the printing composition. The reducing agents convert diazonium salts into inactive compounds. The salts neutralize the alkali and convert the azotol into an insoluble form that is unable to enter the azo-coupling reaction. The metal hydroxide precipitates formed in this process impede the penetration of diazo solutions into the fiber.

To obtain colored reserve colors, a thickened diazole solution is applied to the azotolated fabric in order to completely bind the azotolate in the places where the pattern is applied. During drying, the excess diazole is completely destroyed and when passing the fabric through a diazole solution of a different color, the reaction of azo-combination occurs only in those places where there is no pattern.

Black aniline dyeing or the formation of black aniline on the dye fiber. This black pigment is characterized by its unsurpassed coloristic values and color saturation. It provides very durable colorations that are resistant to light, washing and friction. Dyeing with black aniline *is* rarely used nowadays. However, the possibility of obtaining deep black coloring, unattainable by any other class of dyes, provides its use in direct printing, as well as in obtaining white and colored patterns on the black-aniline background. This pigment is produced directly on the fabric by oxidizing aniline salt ($C H_{65} NH_2 \cdot HCl$) in the presence of catalysts. The dye formation reaction takes place at high temperature and humidity. Potassium chlorate ($KClO_3$) is used as an oxidizing agent. As a catalyst, potassium ferruginous sulfate $K_4 [Fe(CN)_6]$ is used.

In direct printing with black aniline, aniline salt, oxidizer, catalyst, water and thickener are added to the composition of the printing ink. After application of the printing composition, the fabric is dried and treated in a steaming oxidizing burner for 1 minute at a temperature of $98° C$. The fabric is then treated with a K Cr Cr solution. It is then treated with a solution of $K_2 Cr O_{27}$ in the presence of soda or hydrogen peroxide combined with acetic acid to complete the oxidation process. This is followed by a final rinse.

White and colored patterns on a black-aniline background can be obtained only by the *reserve method,* because the coloring is so strong that it cannot be destroyed by etching. Alkaline substances and reducing agents are used as reserve substances. The alkali neutralizes the aniline salt, releasing free aniline, which is not oxidized in a neutral medium. The reducing agents extinguish the action of the oxidizing agents. Reservation is carried out on the fabric impregnated with black-aniline solution containing: aniline salt, $KClO_3$, $K_4 [Fe(CN)_6]$ and water. The material is dried at $60-70° C$ to prevent greenish coloration. A reserve composition containing soda ash, zinc oxide and thickener is printed over the dry fabric. This is followed by drying, passing through an oxidizing burner and washing. In places where the reserve composition is applied, a white pattern is obtained on a black background. Cubic, active dyes and diazaminols can be used to obtain colored patterns on the black-aniline background.

Printing with dispersed dyes. Dispersed dyes are mainly used for printing fabrics made of synthetic and acylcellulose fibers. The mechanism of their fixation on the fiber during dyeing and printing is identical. The correct choice of thickeners is of great importance. Alginates or cellulose and starch esters are used as thickeners. To increase the degree of dye fixation in the composition of printing ink it is recommended to introduce process intensifiers (resorcinol, urea), which contribute to the maximum transfer of dye from the layer of printing ink to the fiber. Fixation of dispersed dyes on hydrophobic thermoplastic fibers takes place in a steam burner at a temperature of $100 - 140° C$ or during heat treatment at $150 - 180° C$ for 4 -5 minutes. Sufficiently effective way of fixation of dispersed dyes on synthetic fibers is the method of transfer thermal printing.

When using high-temperature methods of fixation it is necessary to use dyes resistant to sublimation. At present an effective method of dyes fixation on fabrics from synthetic fibers in printing conditions by short-term (1-3 min) treatment of textile material in vapors of boiling azeotropic mixture of organic solvent and water is proposed. Good results on polyester and polyamide fibers can be achieved by treating the fabric in vapors of azeotropic mixture of benzyl alcohol - water (boiling point 95-100°C) for 2-3 min. The advantage of this mixture is low toxicity of benzyl alcohol and its low content in the mixture (9%). One of the main reasons for the effective action of azeotropic mixtures in fixation of dyes on chemical fibers is the increase in the permeability of the fiber when it is exposed to the fixing medium. When processing printed fabric, the solvent rapidly penetrates into the hydrophobic fiber, forming submicroscopic pores in it and solvating active centers on the surface of these pores. Water vapor enhances the action of the solvent and increases the adsorption capacity of the fiber. In addition, azeotropic mixtures promote molecular dispersion and solvation of dye molecules. All this provides a high rate of diffusion of the dye into the fiber material and the efficiency of azeotropic fixation as a whole.

Questions for knowledge control

1. List the methods of printing with cube dyes. Characterize their advantages and disadvantages. What graduation plates should be used in this or that method of printing fabrics?

2. What chemical processes take place in the sinter when printing with cube dyes by the rongalite-potash method? Why do they use hydrosulfite as reducing agents when dyeing and rongalite as reducing agents when printing?

3. What functions does potash perform when printing with cube dyes by the rongalite-potash method? At what stage of the printing process does potash exhibit alkaline properties and rongalite exhibits reductive properties?

4. What is the fundamental difference between the two-stage printing process with cube dyes and the rongalite-potash process? Which dyeing method imitates the two-stage printing technology?

5. What are the requirements for printing ink thickeners when printing with cube dyes using the two-stage method? What is the role of hydrosulfite, sodium sulfate and alkali in the developer solution?

6. What mills are used for printing of fabrics with cube dyes by rongalite and potash and two-stage methods? Describe the physical and chemical essence of the processes occurring in the ripeners.

7. Give the technology of washing fabrics printed with cube dyes. Characterize the purpose of the processes of fabrics treatment in hydrogen peroxide solution, in boiling soap and soda solution or in the solution of synthetic detergents.

8. The fabric is printed with cube dyes. After processing in the ripper, there was a color change in the dyes. Why? How can the original color of the cube dyes be restored? What chemical processes take place in the sinter and when washing the fabric?

9. List all the operations to which the fabric is subjected when printed with cubozoles by the nitrite method. Explain the mechanism of cubozole manifestation (color recovery of the original cubozole dye) by acid treatment.

10. Explain the purpose of sodium nitrite, glycerol and sodium carbonate in the composition of printing ink. Why does the color of cubosol not match the color of the original cube dye? At what stage of the printing process does the coloration occur? Why?

18 Thickening

In the textile industry, natural polymers such as starch, dextrin, sodium alginate and gum are mainly used as thickeners for printing inks. However, natural thickeners do not always have the full range of properties required for knitted fabrics. In addition, the range of textile products subjected to printing has expanded, the share of synthetic fibers has increased. This has put forward some additional requirements to thickeners, as a result new thickening agents based on modified natural polymers have been obtained. Among thickening agents based on modified polymers, simple and complex starch esters (carboxymethyl and hydroxyethyl), carboxymethyl cellulose (CMC), oxymethyl cellulose, methyl cellulose are widely used. By their printing and technological properties they approach the most qualitative, but expensive and scarce natural thickeners based on traganthus and gum. Another direction of solving the problem of replacing food and expensive thickeners is the creation of synthetic thickeners. Among them polyacrylamide, polyvinyl alcohol, products of alkaline saponification of polyacrylonitrile, salts of polyacrylates have been practically used. However, synthetic thickeners have not yet found wide application, as their printing and technological properties are lower than those of natural polymers.

Thickeners give the printing ink viscosity and stickiness to ensure high print quality: evenness of printing, intensity of coloration of the design, depth of ink penetration into the fabric, clear contours of the design. The thickener must not interact with the dye and other components of the printing ink. It should be easily removed from the fabric when washing. Thickener residue makes the fabric stiff in the areas of the pattern, contains a lot of loose dye. Thickener should be cheap and non-deficient.

High molecular weight thickeners are subdivided into:
- natural;
- artificial;
- synthetic.

These include starch, carboxymethylcellulose (CMC), polyvinyl alcohol, sodium alginate, tragant, carboxymethyl starch (CMC), acrylate thickeners, gum

Colloidal biphasic thickeners are subdivided:
(a) Emulsion;
b) foam

They are two-phase heterogeneous systems. Emulsion thickeners - consist of dispersion medium and dispersed phase in the form of tiny droplets of another liquid. Foam thickeners - consist of a dispersion medium and a dispersed phase in the form of air bubbles. As a dispersed phase in emulsion thickeners use white spirit, velocite, machine or spindle oil. These oils are broken into tiny droplets with a diameter of up to 0.75 microns by mechanical stirring when obtaining an emulsion. The thickening ability of the emulsion depends on the concentration of the dispersed phase and increases with its increase. Surfactants are used as emulsifiers and foaming agents. The disadvantage of emulsion thickeners is flammability and instability during storage.

Table 4
Concentration of thickeners in printing inks, %

natural		artificial		synthetic	
Starch	10-20		6 - 12	Polyvinyl alcohol	

Sodium alginate	5-8	Carboxymethylcellulose (CMC)			30 - 40
Tragant Gum	6 30-50	Carboxymethyl starch (CMC)	8 - 10	Acrylate thickeners	1 - 3

The properties of thickeners and printing inks are divided into two main groups - technical printing properties and physical properties. The printing-technical properties of printing inks include: the evenness of printing, the degree of penetration of printing ink into the thickness of the fabric, the degree of spreading of printing ink on the surface of the fabric, the clarity of the print contours and the intensity of its coloring, the degree of dye fixation, etc. The properties of printing inks are as follows.

Of great importance for the control of the printing process is the establishment of a relationship between the printing and physical properties of printing inks. The lower the viscosity of the printing ink, the deeper the printing ink penetrates into the fabric at the same printing machine speed. The reason for the spreading of pattern contours during the operation of the printing equipment is a decrease in the viscosity of the printing ink due to mechanical destruction of its internal structure. The quality of printing depends on how quickly and completely this internal structure is restored. The property of a thickener to recover its properties after removal of external influences is called thixotropy. If a thickener does not have thixotropic properties, it is unsuitable for use.

The choice of products for thickening printing inks is one of the most important problems in the field of textile printing. The most interesting from the technological point of view are natural substances. Most of the natural thickeners used in the textile industry belong to polysaccharides. In terms of utilization, more than 90% of the natural thickeners used are various starches and products of their modification. Maize (corn) starch is the most commonly used. It is insoluble in cold water and is therefore boiled to make thickeners. Starch is used to make thickeners of 10-20% concentration.

Starch clots have a complex of disadvantages. When stored, they are unstable and quickly delaminate. Printing inks based on them do not give clear contours of the pattern, poorly soak the fabric. Due to poor solubility in water are poorly removed from the fabric when washing, giving it increased rigidity. Starch thickeners can not be used for the preparation of printing inks, which include alkali. Starch coagulates under the action of alkali. The hard film of printing ink formed during the drying of the printed fabric physically prevents the complete transfer of the dye to the fabric. Starch interacts chemically with certain classes of dyes (e.g. active dyes) to form a hard, water insoluble film, which also limits its application. Starch can be modified by heat, acids and mechanically (fine milling). Mechanically processed solvitex starch is used to produce thickeners with high printing properties.

Thermo-starch (pyrodextrins) is formed as a result of short-term exposure of starch to high temperatures (150-200°C). Depending on the exposure time and temperature, products of different degrees of degradation are obtained. They are characterized by different viscosity. Dextrins are products of a deep degree of acid degradation of starch. Dextrin-based thickeners are stable during storage and are resistant to alkaline attack. This makes them suitable for the preparation of printing inks containing a significant amount of alkaline agent (for example, when printing with cube dyes). A disadvantage of dextrins is their low thickening power. Therefore, they are often used in a mixture with starch. Dextrins have adhesive properties. This allows them to be used as an adhesive in photographic film printing.

Modification of starch by its esterification, i.e. obtaining soluble starch esters. The most common thickener is carboxymethyl starch, which is obtained by the action of monochloroacetic acid on starch treated with alkali.

$$Крахм - OH + NaON \rightarrow Крахм - ONa + H_2O;$$
$$Крахм - ONa + CH_2Cl\,COOH \rightarrow Крахм - OCH_2COONa + HCl$$

KMC is recommended for applications where a viscous thickener with a low dry matter content is required. Printing inks based on KMC thickener allow reproduction of large areas with one color on fabric. KMC thickener is stable during storage and has no reducing properties. It is 2-3 times more expensive than starch.

Oxyethylated cellulose derivatives (OECs) and cellulose ethers. Carboxymethyl cellulose (CMC) is used in the form of sodium salts. Compared to starch-based thickeners, CMC-based thickeners reduce the color intensity of the fabric. The reason for this is the low structured nature of CMC-based printing inks, which promotes deep penetration of the printing ink into the fabric. This leads to lower surface coloration of the fabric. CMC has a sufficiently high thickening capacity already at 6% concentration. It is well soluble in cold water. When the temperature rises above 50°C, its solubility decreases significantly. Due to this, printing inks based on CMC do not spread out and do not stick when steaming and wet processing of the printed fabric. CMC-based thickeners are stable during storage and can be easily washed out of the fabric with cold water.

One of the sources of thickeners is sea brown algae. This algae contains alginic acid, the sodium salt of which is used as a thickener called sodium alginate.

Sodium alginate is a hard and brittle dark-colored laminae. It is easily soluble in water and at low concentrations gives a homogeneous and viscous mass. A 8% solution of sodium alginate is commonly used. Sodium alginate thickeners are resistant to alkalis and unstable to acids. They have good covering power and are easily washed out of the fabric when rinsed. Sodium alginate is compatible with all anionic dyes, pigments and metal complex dyes. It is incompatible with cationic dyes. Abroad alginates for textile industry are produced under different names: manutex, lamitex, etc. Manutex alginates contain 10% calcium alginate and have a very high thickening ability. Manutexes are used for printing with direct, metal complex, disperse, acid, active, cube dyes.

Table 5
Assortment of thickeners for printing with active dyes

Thickener	Chemical nature	Thickening mode	Properties of thickeners and thickeners
Sodium alginate	Sodium salt of alginic acid derived from seaweed	Thickening is prepared by hot method at cooking temperature of 80-85° C for 3-5 hours	The plates are greenish-brown in color, stable in the pH range of 4-10, anionic. The thickeners have good covering power The following table provides a list of the

			most important factors to be aware of
Manutex RS-92; RS-21P; RS-23P; RS-26P; RS-32P, F-560; M-330.	Sodium salt of alginic acid containing Ca ions^{2+} , Mg ions^{2+} , and additives of electrolytes, urea, etc. for various grades.	Thickening is prepared by cold or hot method at cooking temperature 60-80° C for 30 min. Depending on the water hardness and to regulate the flowability of the thickener, complexing agents up to 10 g/kg of thickener are added.	Powder from yellowish to greenish-brown color, fully soluble in water, stable in the pH range 4.5-11.0 , anionic, sharp gelation with polyvalent metal salts is observed. The thickeners form elastic films, easily removed by washing after treatment under any temperature conditions of fixation.
Solvitose P-5	Carboxymethyl ester of starch	Preparation by cold method with high-speed stirrer or by hot method at a temperature of 50-60° C for 60 minutes	White flakes, completely soluble in water, pH 9-10, anionic type, sensitive to polyvalent metal salts, form hard films on fabrics at high-temperature treatments that are difficult to dissolve
Polyprints L-390, L-400.	Sodium salt of alginic acid	Thickening is prepared by cold or hot method at cooking temperature of 60-80° C for 30 minutes	Medium (L-390) and low (L-400) viscosity thickeners for high quality printing and fine contouring of cellulose fiber textiles
SNT-Alginate SMT	Sodium salt of alginic acid	Thickening is prepared by cold or hot method at a cooking temperature of 60-80° C for 30 min.	High viscosity thickener with very high leachability and versatility in application.

Natural gum - is a solidified sap protruding from the bark of some species of tropical plants. Externally, they are pieces of different sizes, fragile and horny. According to the chemical structure of gum is a mixture of polysaccharides. They have a number of valuable properties that make them indispensable for etching and reserve printing. These products are of great importance when printing on fabrics made of synthetic fibers. 50% solutions are used. The disadvantage of gum is its very high cost and ability to interact with the dye. This leads to a noticeable brightening of the color, contamination of the background of the fabric and loss of dye. *Tragant* (fruit gum) - has a good thickening ability. It is prepared in 6-8% concentration Tragant gum is used in a mixture with starch or dextrin for printing on fabrics made of cellulose fibers.

Synthetic thickeners - use mainly carboxylic linear polymers:
- polyacrylamide;
- saponification products of polyacrylonitrile;
- polyvinyl alcohol;
- polymers of acrylic and methacrylic acids.

Colloidal two-phase thickeners. In such thickeners, one component (inner phase) is uniformly distributed in the volume of the second component (outer phase). Their common feature is the presence of a low molecular weight thickener. The outer phase of such thickeners must necessarily be liquid. The inner phase can be in three states: solid, liquid and gaseous. If the thickener is solid particles, the thickener is called suspension, if the thickener is a liquid (oil) - emulsion and if the thickener is a gas (air) - foam. Emulsion thickeners are currently the most used.

The thickening capacity of an emulsion thickener depends on the concentration of the liquid thickener (internal phase) in it. The maximum thickening ability is achieved at a concentration of about 90%. With further increase of concentration the emulsion stability decreases and it stratifies. Emulsion thickeners are most common in pigment printing, where the presence of high molecular weight thickeners is undesirable. Emulsion thickeners are obtained by mechanical mixing of oil with an aqueous solution of emulsifier. In this case, oil droplets are distributed in the aqueous medium, forming spherical particles up to 0.75 microns in size. The disadvantage of emulsion thickeners is their flammability and low stability in the presence of a large amount of electrolyte.

Thickeners and printing inks are prepared in the dyehouse of the printing shop. The dyehouse must have a warehouse for dyes and chemical materials, a dyeing room, a room for cooking thickeners and preparing printing inks, a warehouse for storing inventory and utensils, and a room for the shop laboratory. Water used for cooking thickeners and dissolving dyes, must be softened, clean and transparent. For this purpose it is possible to use condensation water, which saves steam used for water heating.

The operation of the inksetter is based on compliance with the exact weight and volume ratios of the constituent parts of the printing ink as specified in the recipe, which can be accomplished if the inksetter has the appropriate equipment. All constituent parts of the printing ink should be carefully weighed, volumetric measurements are acceptable for liquids with a specific gravity equal to one. The accuracy of weighing in the ink cooker should be high, as deviations in weight can lead to changes in color intensity, changes in the consistency of the printing ink and other properties, to overconsumption of chemical materials and to an increase in the cost of production.

For the preparation of thickeners that require heating and stirring, it is necessary to have special apparatuses - open boilers and autoclaves. Open boilers have double walls: external (cast iron) and internal (copper or stainless steel). Sharp steam or water enters the space formed by the boiler walls. The boilers should be rotated in a horizontal plane to facilitate pouring out of them ready thickeners. For thorough and continuous mixing of the contents of the boiler in it set one or two copper mechanical stirrers removed from the boiler. For ease of service boilers are installed in a row, forming the so-called battery, containing boilers from 25 to 400 liters. Boilers of small capacity are designed for boiling dry dyes. For hard-to-digest thickeners (tragant, gum, etc.) is necessary to have a boiler-autoclave, working under pressure. Ready thickening from the tank for cooking pumped by a circulating pump in the receiving tank. Duration of cooking is about 2.5 hours. During this time about 1000 kg of thickening can be prepared. The creation of flow lines for the preparation of thickeners and printing inks is increasingly used, for example, created an automatic plant UPK-P2 with program control for the preparation of printing inks. In one cycle this unit can prepare up to 50 kg of ink.

To mechanize the process of mixing the components of printing inks use speed agitators of various types (screw, paddle, etc.), equipped with an electric motor with an elongated shaft, at the end of which the agitators are fixed. Turbine stirrers single or double are also used. High-speed agitators can be installed in different ways: by hanging, fixing on the wall with a bracket that allows it to move in the vertical direction, fixing clamps on the edge of the bowl, installation on a mobile cart, etc., and so on. High-speed stirrers are used for preparation of various emulsions (of the "oil-in-water" or "water-in-oil" type). By means of intensive stirring a good emulsification of both phases is achieved. Necessary equipment of the paint shop are also paint grinders, colloidal and roller mills, which serve for thorough mixing of printing inks, preparation of overlapping whites, etc. The equipment of the paint shop is also necessary.

Equipment designed to free thickeners and printing inks from possible mechanical solids that could damage the equipment during printing belongs to a separate group. The simplest is the manual method of filtering - rubbing the thickener with brushes through a cloth or sieve - used for small quantities of printing inks or thickeners. For filtering large quantities, mechanical brush tedders are used, as well as vacuum tedders, which give high productivity. Dyeing machine should have apparatus and machines for washing dirty dishes and other inventory. For smooth operation in the paint shop should be a sufficient number of auxiliary equipment - tubs, tanks of different capacities, buckets, sieves, wooden stirrers and others. Ready printing inks are delivered to the place of consumption in stainless steel tubs or other containers.

19 Printing Technology

Technology of printing cotton and viscose staple fabrics with active dyes by one-stage method

Composition of printing ink, g/kg:

Active colorant 20-80
Water (70-80°C) 150-200
Sodium hydrogen carbonate (or sodium carbonate) 10-20
Ludigol 10
Urea 100-120 (for steaming methods)
150-200 (for thermofix. methods)
Thickener (alginate, synthetic, based on
carboxymethyl starch or mixed) up to 1000 g.

The order of printing on the machine with rotary grid templates (f. "Stork"):
1. Application of printing ink.
2. Drying fabric in a print dryer.
3. Fixing the dye by one of the methods:

(a) Steam method in saturated water vapor at 102-105°C for 3-8 min (depending on the reactivity of the dye and the area of the pattern) in curtain or reduction mills;

b) thermofixation method:

c) *hot air* at 150-170°C for 5-3 min on heat treatment machines of MBRT type;

d) IR-radiation at the tissue surface temperature of 180-190°C for 15-10 s at the line LT-180.

4. Washing and drying of fabric on washing and drying machines, for example, LPS-180-12.

Processing in the wash part of the line:

The 1st bath is cold water;

2nd bath - hot water at 85-95°C;

3rd bath - solution of non-ionogenic CMC (1.5-2 g/l) at 90-95°C;

The 4th and 5th baths are hot water at 55-65°C;

The 6th bath is warm water at 45-55°C.

5. Drying of fabric on the drying-drum machine of type MSB 2-3/180.

Technology of printing natural silk fabrics with active dyes

Composition of printing ink, g/kg:

Active colorant 10-80
Water (70-80°C) 200
Urea 100
Sodium hydrogen carbonate 10-15
Ludigol 10
Alginate thickener up to 1000

The *order of printing* on the machine with flat grid templates (Meccanotessile, Stork "Rigiani Macchine".
1. Application of printing ink.
2. Drying in a print dryer.
3. Stacking fabric in the cart.

4. tension-free processing in a curtain-type ripper (Arioli, Stork, Kyoto) at 102-104°C for 25 min.
5. Washing of fabric on the machine type MKP-1 according to the mode:
with urea solution at 15-20°C for 15-20 min;
with cold running water for 40-60 minutes;
with a solution of non-ionogenic CMC (≈2g/l) at 60-65°C for 60 min;
with softened water at 40-45°C for 15 min;
with a solution of non-ionogenic CMC (≈2g/l) at 85-90°C for 60 min;
with softened water at 55-60°C for 20 min;
in a solution of 25% ammonia (1 g/L) and hexametaphosphate (1 g/L) at 45-50°C for 20 min;
with cold running water for 20 minutes;
revitalization in acetic acid solution (≈1,5-2g/l) at temperature 20-25 ° C for 15 min;
6. Drying the fabric on a tumble dryer at 110-115 C.°

Technology of printing wool fabrics with active dyes

Printing ink composition, g/kg:

Active dye (labeled "SH" or "T", Lanazoles, ostazines with "H" mark)	2-25
Urea	20-100
Water (70-80°C)	150-200
Thickener (sodium alginate and non-ionogenic guar derivatives)	up to 1000

The *order of printing* on machines with flat mesh templates (Meccanotessile, Buser, Righiani, Zimmer, Stork).
1. *Drying* in a print dryer.
2. *Steaming*:
a) on batch equipment (for piece products) with saturated water vapor at 102-104°C for 65-75 min;
b) on continuous equipment (Arioli milling machine) at a temperature of 100-102°C for 40-90 minutes (depending on the area of the pattern).
3. *Washing of* fabrics on the unit of two lines LZP-180 Sh.
1-2nd baths - water (30-35°C);
3-4th baths - ammonia solution (pH 8-8.5) at 80°C;
5th-6th baths - CMC solution (0.1-0.2 g/l) at 50-60°C;
7th-8th baths - water (30-35°C);
9th-14th baths - water (25-30°C);
15th-16th baths - solution of 30% acetic acid (1g/l) at 25-30°C.
4. *Drying* in a drying machine at 85-95°C.
Disperse dyes. Dispersed dyes belong to the class of non-ionic dyes. Most of them are azo dyes, derivatives of anthraquinone and nitrodiphenylamine, possessing molecules of small size and uncomplicated structure (relative molecular weight 250 - 300). For sublistatic transfer printing special release forms of highly concentrated disperse dyes from among low-energy, i.e. easily sublimable dyes are used. Of the large assortment of disperse dyes, direct printing usually uses those dyes that have a relatively high lightfastness and stability at high temperatures.

Table 6
Disperse dyes for printing

Manufacturer	Title
Hunstman (Ciba)	**Terazyl P**
	Powdered
	Yellow P-6GS
	Golden yellow P-4R.
	Orange P-GL
	Brown P-3R
	Red P-3G
	Red P-4GN
	Pink P-2B
	Red P-3BS
	Ruby P-2G
	Red P-4BS
	Purple P-BL
	Blue P-2BR
	Blue P-BGE
	Blue P-BS
	Navy blue P-GRL
	Black P-GR
	Liquid
	Yellow P-6GS 25%
	Golden yellow P-4R. 50%
	Orange P-GL 50%
	Brown P-3R 50%
	Red P-3G 50%
	Red P-4GN 75%
	Pink P-2B 100%
	Red P-3BS 100%
	Ruby P-2G 50%
	Red P-4BS 50%
	Purple P-BL 100%
	Blue P-2BR liquid. 50%
	Blue P-BGE liquid. 67%
	Blue P-BS 67%
	Navy blue P-GRL 67%
	Black P-GR 67%
	Terazyl X
	Yellow X-5G liquid.
	Red X-GSA liquid.
	Blue X-HLB liquid.
	Gray X-SGN liquid.
	Yellow X-GWL liquid.
KRATA Tambov NPO	**Dispersed**
	Yellow 4 Z 200%
	Yellow 2 Z 200%

	Yellow 75%, 100%
	Golden yellow
	Orange G
	Orange 2K
	Yellow-brown 2G
	100%, 150%
	Scarlet 2J
	Scarlet
	Scarlet C
	Pink 2C 200%
	Ruby G
	Ruby 100%, 200%
	Red-brown G
	Brown 150%
	Dark brown 2J
	Turquoise 200%
	Blue Z 200%
	Blue 2 Z 200%
	Blue 2 100, 150%
	Dark blue Z 100%,200%
	Dark blue 75
	Dark blue ST 300%
	Purple
	Dark green 2G
	Gray
	Black 2K 100%, 150%
	Black Z
	Black CH-1
	Black KC 150%
	Black ST 300%

In addition to the dye, printing ink usually includes urea or dicyandiamide, a non-volatile acid-donating agent (more often oxalic acid mixed with ammonium sulfate), a weak oxidizing agent, and a thickener.

There are no special requirements for thickeners for printing with disperse colors. Thickeners based on sodium alginate (e.g. Lamalgin GS 5), esterified starch (e.g. Emprint CE) or their mixtures (e.g. Prisulon CM 5-10) as well as synthetic thickeners, which have been widely used in recent years, are used for thickening printing inks.

Hunstman (Ciba) offers a synthetic thickener for printing disperse dyes on polyester textiles, Lioprint DTCS. It is a dispersion of acrylic polymer of anionic nature. It is characterized by stability to hard water and alkalis. It may precipitate in strongly acidic environments. The thickener is compatible with all anionic and non-ionogenic TVBs used in printing. The use of Lioprint DT-CS Thickener provides:
- greater color output;
- significantly better contour definition and evenness than most natural thickeners;
- achieving high homogeneity of the printing compound while reducing the preparation time.

This thickener is not affected by bacteria, hence the prepared thickener can be stored for a long time maintaining its original viscosity without the addition of biocidal additives. Lioprint DT-CS can be used in combination with natural thickeners. Even relatively low concentrations of synthetic thickener (e.g. 10 g/kg) mixed with natural thickeners significantly improve color yield and print evenness. When printing on cotton-polyester fabrics with active and dispersed dyes, Lioprint RDHT thickener has proven to be the most effective.

Printing of fabrics and knitwear made of polyester, polyamide and acetate fibers with disperse dyes. Dispersed dyes belong to the class of non-ionic dyes. Most of them are azo dyes, derivatives of anthraquinone and nitrodiphenylamine, possessing molecules of small size and simple structure. Printing is carried out by direct printing or by the "sublistatic" method. Direct printing with disperse dyes of textile materials of different fiber composition differs only by the used equipment for fixation and temperature and time parameters of the process. Usually, direct printing with disperse dyes uses printing compositions based on natural or synthetic thickeners.

Composition based on natural thickeners, g/kg

Dispersed dye	5 - 30
Deaerator (e.g. LYOPRINT AIR)	3 - 5
Acid/acid donor (e.g. citric acid)	up to pH 5 - 5.5
Oxidizing agent (e.g. sodium chlorate)	10 - 20
Fixation gas pedal	0 - 20
Natural thickener (sodium alginate pure or mixed)	500 - 600
Water	up to 1000

Composition based on synthetic thickeners, g/kg

Dispersed dye	5 - 30
Deaerator (e.g. LYOPRINT AIR)	3 - 5
Synthetic thickener (LYOPRINTDT-CS)	35 - 45
Water up to	1000

When synthetic thickeners are used, the process of preparing printing compounds is greatly simplified. Oxidizing agent, fixation gas pedal and acid donor are usually not required in this case, as the synthetic thickener fulfills the functions of these substances. The preparation of the printing compound involves only the addition of dye and water.

Dispersed dyes are of particular interest for coloring fiber blend fabrics. When printing on fabrics from a mixture of fibers, dyes of different classes are combined to color the fibrous components of the fabric. Combinations of dispersed and active dyes are widely used for printing of materials from polyester fiber and cotton blends. The following requirements should be taken into account when composing dye mixtures of the mentioned classes:

- the closeness in tone of coloration on the cellulose and polyester components of the fabric;
- No interaction between active and dispersed dyes in the printing ink and on the fiber during the fixation process;
- high sublimation stability of dispersed dyes;
- low sensitivity of dispersed dyes to alkaline medium;
- possibility of using alkaline media at the fixation stage and at the washing stage.

Of particular importance is the washing of material printed with a mixture of dispersed and active dyes. The complexity of the washing process is due to the fact that the optimum temperature conditions for removal of unfixed dispersed and active dyes are different. While removal of active dyes requires a wash water temperature close to boiling, for dispersed dyes the optimum wash temperature is 50 - 60°C. At high temperature, and especially in the presence of urea, the presence of which in the printing ink is necessary, the latter stains the cellulose fiber, forming a defect - painted white background of the fabric. In this regard, the washing process is carried out in such a way that first the dispersed dye is removed, and then the active dye. First, the washing is carried out with cold running water, then with warm water, followed by washing first at a temperature of 40°C, then at a temperature of 60°C. The temperature of the washing solution is then increased to remove any unfixed active dye. Washing is carried out on pass-through apparatuses consisting of 8 - 9 washing boxes, it is advisable to wash twice: the first - to remove dispersed dye, the second - to remove the active dye.

When printing cotton-polyester fabrics with mixtures of dispersed and active dyes, the most complete transfer of both dyes from the printing ink to the fiber components of the fabric is achieved when sodium alginate is used as a thickener.

Technology of printing polyester fabrics and knitted fabrics with dispersed materials
dyes

Composition of printing ink, g/kg:

Dispersed dye	5-35
Acid-bearing agent (oxalic, acetic acids)	up to pH 5-6
Oxidizer (ludigol)	10
Fixation Accelerator (Rapidprint RE)	15
SNT-Defoamer BS	1
Mother clot from Prisulon SM 5-10	up to 1000

The order of printing on the machine with Stork grid templates:

1. *Drying* in a print dryer at 120-130°C
2. *Ripening* with superheated steam in an Arioli r*ipper* at 175-180°C for 7-6 min.
3. *Washing and drying*
(a) Washing and drying of fabrics on the Kleinevefers line:
soak in impregnation box with water (T=15-25°C);
rinsing with water sprays (T=15-25°C);
washing in the washing part of the line:
1 bath - warm water (T=45-50°C);
2 bath - alkaline reducing solution containing sodium dithionite -2.0 g/l, sodium hydroxide - 2.0 g/l and CMC - 0.5 g/l (T=55-60°C);
relaxation bath - a solution containing sodium hydroxide - 3.0 g/l and CMC - 1.5 g/l (T=70-75°C);
3 bath - hot water (T=55-65°C);
4 bath - warm water (T=50-55°C);
5 bath - cold water (T=20-30°C);
4. *Drying* in a drying chamber.
b) *Washing of knitted fabrics* **on the** Arioli line: locking on a perforated drum (T=15-25°C);
flushing:
1 bath - solution containing CMC - 1.0 g/l and sodium hydroxide - 1.5 g/l (T=55-60°C);
relaxation bath - alkaline-reduction solution containing sodium dithionite - 2.0 g/l, sodium hydroxide - 3.0 g/l and CMC - 1.5 g/l (T=60-65°C);
2 bath - water (T=55-60°C);
3 bath - water (T=20-30°C);

5. Drying on Brückner SSM at temperature up to 140°C.

Questions for knowledge control

1 Describe two basic schemes for printing cellulose fiber fabrics with active dyes. Characterize the purpose of the alkaline agent in the composition of printing ink. Can sodium hydroxide be used as an alkalizing agent?

2 Identify the role of urea in the composition of printing ink when printing with active dyes. In which printing method is it more necessary - steam or thermal?

3 Identify the principle of selecting thickeners for printing with active dyes. Can starch be used as a thickener? Justify the answer.

4. What are the requirements for disperse colorants when used in printing? Give the composition of the printing ink and characterize the role of TVBs included in it.

5. Under what conditions is the fixation of disperse dye on the fabric? Which method of fixation is more effective? Which method will you choose for polyester fiber fabrics?

6. Give the mechanism and technology of the transfer thermal printing method. Why can this method be used to print patterns on piece goods and on the cut parts of garments?

7. What dyes can be used for heat transfer printing? What fibers can be printed with this method?

8. What are the requirements for the substrate paper used in thermal printing? What factors affect the rate of dye transfer from paper to fiber?

9. Characterize the advantages and disadvantages of the method of transfer thermal printing. Why this method is not widely used in cotton and wool textile industries?

20 Printing with flat mesh templates

Printing of fabrics with mesh templates is done on printing tables.

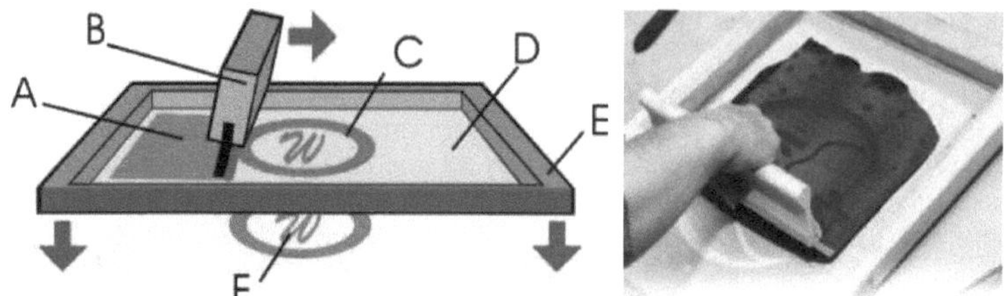

A- printing ink; B- rakla; C- pattern on mesh template; D- mesh template; E- frame; F- printed pattern on fabric.

Figure 36. Principle of fabric printing with mesh templates

The fabric prepared for printing is spread out along the printing table, well spread across the width and pinned along the edges with special pins. Printing of fabrics on manual tables is carried out by wiping the printing ink through the template with the help of a rakley.

Figure 37. Wiping printing ink through a template

Mesh pattern printing is mainly used for printing silk, wool and knitted fabrics, which are easily deformed by mechanical effects on the fabric. The principle of printing is that a pattern is applied to the fabric, evenly spread out and attached to the elastic pad of the printing table, by wiping or pushing the printing ink with a rakley through the mesh template. The great advantage of flat mesh template machines is that changing the number of templates does not affect productivity, as all templates are lowered and raised at the same time. The cost of making a set of printing templates is much lower than the cost of making a set of engraved rolls to reproduce the same pattern. Printing with flat mesh templates is expedient when the circulation of the drawing is small. This method can be used to reproduce almost all types of patterns (except for longitudinal stripes) and any degree of complexity.

Figure 38. Using flat templates to print fabric

The templates can be flat rectangular in shape if the mesh is stretched over a rigid frame, or cylindrical if the mesh is in the form of a tube. With this in mind, there are *3 types of machines* with mesh templates:
-magazine;
- with stationary templates;
- with cylindrical templates.

The shop-type machine is a short conveyor belt with a length of one pattern pattern pattern. The fabric is glued to the conveyor belt to avoid shifting of the finishing elements of the pattern. Patterns with paint come from a special store one after another. After applying paint to the fabric, the conveyor belt is moved to the size of one template and the process is repeated. The equipment is small-sized, but has a high complexity of construction and low productivity.

Machine with *stationary templates - a* conveyor over which templates (up to 8 pcs.) are permanently installed.

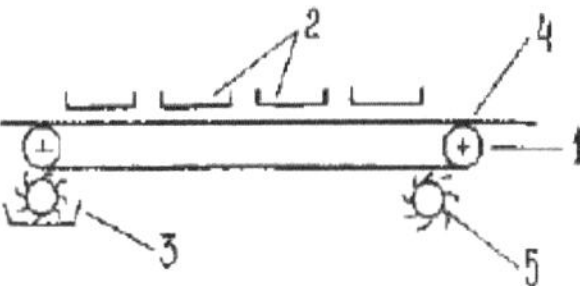

1 - conveyor; 2 - templates; 3 - trough with glue; 4 - fabric; 5 - brush for cleaning the conveyor.

Figure 39. Schematic of the machine with stationary templates

Printing ink of different colors is fed into each of them and rakles are simultaneously switched on in the templates. After the ink has been wiped off, the templates are lifted and the conveyor belt with the fabric is moved by one rapport. The process is then repeated. Rackles are made of rubber, resistant to chemical reagents. The machines are indispensable for printing multicolor patterns of increased complexity on fabrics with easily deformable structure, but they have significant dimensions and insufficient productivity.

Cylindrical pattern machines are a conveyor belt with seamless, thin-walled, perforated metal mesh cylinders above it, acting as a printing roll. This method allows the printing of designs on a wide range of products, from carpets and floor coverings to the finest fabrics, knitwear and paper.

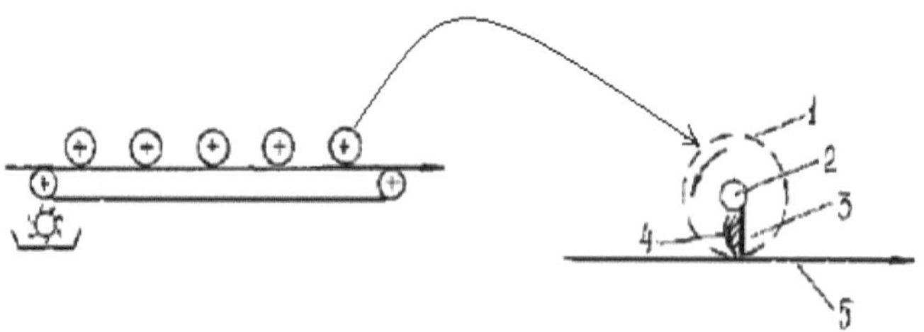

1 - grid; 2 - tube for ink supply; 3 - rakla; 4 - printing ink; 5 - fabric

Figure 40. Machine with cylindrical templates

The template is a perforated nickel cylinder. It is produced by galvanic method. The pattern is transferred to the template by photochemical method. Printing machines with cylindrical grid templates have a simpler design than machines with flat templates. They have higher productivity. The average speed of the machine is 45-60 m/min. When working on machines with cylindrical templates exclude such types of defects such as "drags" (smearing) paint, distortion of the pattern. The racket is made in the form of thin steel or rubber blades, located at an angle to the template and connected to a device for metered supply of paint inside each template. The machines are highly productive, working in a continuous mode. They are expensive due to the difficulty of manufacturing the templates.

21 Mechanism of the printing process from a screen plate

The mechanism of the printing process from a stencil plate can be conventionally represented as follows.

When the doctor presses the printing plate against the material to be printed, each printing element forms a space bounded at the bottom by the material to be printed and at the sides by the printing elements. Ink moved by the doctor blade over the plate fills the space of the printing element, forming an image directly on the material to be printed. As the doctor passes over the printing element, the ink from above is cut off by its working edge. When the printing plate is retracted, the filaments of the screen are extracted from the ink adhering to the material to be printed.

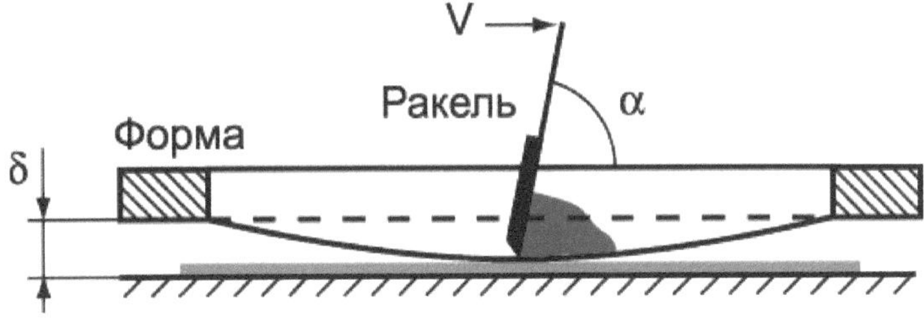

Figure 41. Diagram of the stencil printing device

Four stages can be distinguished in the process of forming an ink image on a print: creation of the space of the printing element, filling it with ink, withdrawal of the printing plate from the material to be printed, and fixation of the ink image on the print. The quality of the colorful image formed in this way depends on the quality of the space of the printing element, the degree to which it is filled with ink, the conditions of interaction between the ink and the plate and the material to be printed, as well as on the structural and mechanical properties of the ink. In screen printing the quality of the space of the printing element depends on the evenness of its contour, microgeometry of the contacting surfaces of the printing plate and the material to be printed, as well as on the density of their mutual contact at the time of formation of the colorful image on the print. The amount of ink pressed through the grid cells is determined by the size of the space of the printing element, the viscosity of the ink, the pressure acting on it, and the time of pressure action. The required image quality on the print can be achieved if the amount of ink that passes through the printing elements of the grid during a printing cycle strictly corresponds to the volume of the space of the printing element.

Factors affecting the formation of the ink layer on a print:
1. characterization of the applied mold base mesh;
2. A method of manufacturing a printing plate;
3. the nature of the material to be sealed;
4. Rheological properties of paint;
5. Rackel hardness and edge width;
6. Modes for carrying out the printing process;
7. Distance between the mold and the material to be sealed;
8. Raquel angle and pressure;
9. The amount of ink remaining on the screen after the printing plate is retracted.

These factors are largely calculable and manageable over a fairly wide range.

If, when printing, the doctor presses the plate tightly against the material to be printed and slides on top of it, the volume of the printing element space V will be:

$$V = S (h + h + h_{123}) - Vc; \qquad (2)$$

where S is the area of the printing element;
h , h , h_{123} - respectively thickness of the grid, thickness of the copying layer under the grid, average height of microroughnesses of the material being sealed;
V_c is the volume occupied by the filaments of the mesh inside the printing element.

When ink fills the entire volume of the printing element space, the thickness of the ink layer formed on the print does not always correspond to the calculated volume, i.e. part of the ink remains on the grid after the plate is retracted. The amount of ink transferred to the print depends on the viscosity of the ink and the angle of inclination of the doctor. In real conditions, there may be variants when hydrodynamic pressure arising in the ink at small angles of inclination of the doctor becomes greater than the pressing force of the doctor made of elastic material, bends its edge and raises the doctor above the mold. In this case, the actual thickness of the formed ink layer will be greater than the space volume of the printing element. The volume of ink $V_\text{к}$, transferred to the impression, will be different from the volume of the printing element space and will be equal to:

$$V_\text{к} = [S(h1 + h2 + h3 + h4 + h5)] - (V_c + V_o), \qquad (3)$$

where S is the area of the printing element;
h1,...,h5 - respectively the thickness of the grid, the thickness of the copying layer under the grid, the average height of microroughnesses of the material to be sealed, the height of the doctor lift under the action of hydrodynamic pressure, the gap between the mold and the surface of the material to be sealed;
V_c is the volume occupied by the filaments of the mesh inside the printing element;
V_o is the volumetric amount of paint remaining on the grid after mold removal.

The transfer coefficient ($K_\text{п}$), calculated as the ratio of the volumetric amount of ink on the print ($V_\text{к}$) to the original amount of ink, determined from the volume of the printing plate space ($V_{\text{п.з}}$), may be less than, equal to, or greater than one:

$$K_\text{п} = V_\text{к} / V_{\text{п.з}} \quad (V_\text{к} < 1, V_{\text{п.з}} > 1) \qquad (4)$$

At present, the most widespread type of photographic film printing machines with flat templates are the machines of Meccanotessile (Italy), Stork (Netherlands), Reggiani McKinet (Italy), Zimmer (Austria), Buser (Switzerland). Basically all machines are equipped with missiles in the form of double rubber strips and only machines of "Zimmer" (Austria) and SACM (France) are equipped with magnetic missiles in the form of metal rods of different diameters. The use of machines with flat mesh templates is most appropriate when printing small batches (up to 4000 m) of silk, woolen fabrics or knitwear, as well as for the design of piece goods made of other types of fibers (handkerchiefs, napkins, etc.) of small circulation of the pattern.

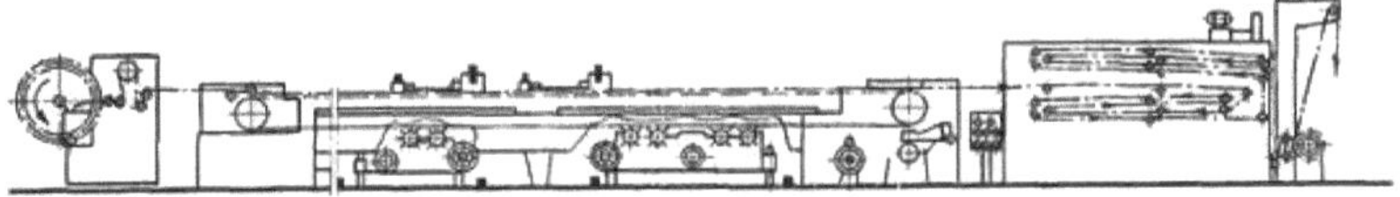

Figure 42. Diagram of the printing machine with flat mesh templates

Printing machine with flat templates of the company "Meccanotessile" (Italy). Composition:
- filling device - an insertion rack with a roll of fabric and a system of rollers that feeds the fabric onto the endless web of the printing press;
- adhesive device for water-soluble and permanent adhesive. The water-soluble adhesive is applied continuously with a special brush and two adhesive adhesives, one of which provides the required thickness of the adhesive layer on the print mat. When using the permanent adhesive system, the adhesive is pre-applied to the carefully degreased endless web of the printing press by means of a special metal rake installed at the machine inlet;
- a device for drying the endless web of the printing press with hot air (65 - 95° C) supplied by a fan;
- a device for washing the endless web of the machine with two pairs of brushes;
- 18 m long padding table with an endless web moving in horizontal guides. The web movement is intermittent for the length of the set report with an accuracy of ± 0.1 mm; when applying permanent adhesive, the web movement is continuous. The web movement speed is 5 - 20 m/min;
- a rake assembly that provides transverse movement of rakes in templates installed in template holders fixed on special slides. Lifting of templates is oblique. Racquets are double rubber open type. The pressure of the rake is changed by adjusting the height of its installation, pressing force and degree of sharpening;
- 7.4 m long drying chamber with direct or overhead input of fabric from an endless web. In the chamber, the fabric first travels on a conveyor, then on guide rollers and enters the selector;
- sampling device, which is a self-laying device with an adjustable pneumatic compensator that ensures constant fabric tension in the chamber.

Printing machine with flat templates by "Stork" (Netherlands).
Composition:
- filling device for transporting fabric with adjustable tension on guide rollers from a roll up to 1000 mm in diameter with centering inlets feeding the fabric to the stuffing table through a curvilinear heated plate mounted vertically in front of the stuffing table;
- adhesive device for applying permanent adhesive of thermoplastic type to the web of the printing machine, which is a metal knife with side stops adjustable to the required printing width and manual adjustment of the thickness of the applied layer. For better pressing of the fabric to the adhesive layer there is a special pressure roller with pneumatic pressure;
- An apparatus for washing an endless web that cleans the web from contaminants after printing by means of a pre-wash comprising a spray tube and a polyester fabric frame, and a wash with a washing machine comprising a stainless steel tank with two tubular sprinklers and two rotating brushes;

- a device for drying the endless web, consisting of special scraper racks that dry the web. Both devices are equipped with a pneumatic lifting and lowering system;
- 16 m long stuffing table with an endless web moving in horizontal guides by means of a hydroelectric system with a rapport accuracy of ±0.1 mm.

To create elasticity, the table is covered with a 10 mm thick felt fabric and a wear-resistant polyester fabric on top of it. Adjustment of the pattern is stepless. The speed of web movement is 4 - 20 m/min. Double open-type rubber slings moving in the transverse direction. Devices for installation and fixing of templates provide oblique lifting and lowering of templates and movement of racks. The drive of rakley and templates and their lifting are pneumatic. The endless web with glue, on which the fabric is supplied, is heated by infrared emitters, automatically switched off at the moment of contact of the fabric with the adhesive coating and switched off when the movement of the endless web stops;
- 4.1 m long conveyor-type nozzle drying chamber with polyester belt conveyor and upper fabric inlet;
- sampling device - self-laying with fabric stacking in a cart, equipped with an automatic speed control device.

Printing machine with flat templates by "Reggiani Macchine" (Italy). Composition:
- filling device equipped with a roller compensator and centering glands;
- adhesive device of two types: for continuous application of water-soluble adhesive with regulation of the thickness of the applied adhesive layer and temperature of the adhesive and for application of permanent adhesive;
- rubber web washing device with three rotating brushes and a web drying device with a fan and steam radiators;
- 16 m long stuffing table with an endless printing web, which is a high-stiffness rubber web with a special synthetic elastomeric coating and the actual printing table with an elastic coating. The endless web is moved by a hydropneumatic drive with a high accuracy of ±0.1 mm. Templates are fixed on the table with the help of special devices with autonomous pneumatic lifting and lowering. Template centering is ensured by special adjusting handwheels, and the movement of the raclis is provided by a hydraulic drive. Dual rakli are used, their angle of inclination and speed of movement are adjustable within wide limits. Racquets are made in the form of a rubber blade, one type of rake is used, the profile is changed by the angle of sharpening, the height and angle of inclination of the rake can be changed. Fastening of the rakley at intermediate points;
- 7.1 m long conveyor-type drying chamber with upper fabric input into the chamber;
- selective device - self-laying for laying fabric "in a book".

***Hydromag* printing machines (*Boozer* Switzerland).** Swiss company *"Buser"* created several models of machines for printing flat mesh templates, known under the brand name *Hydromag 3-6.* The most perfect of this family of machines is Hydromag-6, designed for printing fabrics and knitted fabrics with easily deformable structure.

A distinctive feature of this model of machine is that the lower part of the conveyor belt, which is located under the printing table and in the filling unit, moves continuously, whereas on the surface of the printing table it normally makes a translational movement with an amplitude determined by the pattern pattern pattern. In order to ensure perfect coordination of the translational and continuous movement of the transport belt, the filling device has a guide roller working on the principle of a compensating roller, i.e. making movements with a large amplitude in the vertical plane. Similar pendulum movements, only in the horizontal plane, are made by a guide roller located under the printing table. This unique in printing conveyor belt drive system ensures extremely uniform, continuous and virtually tension-free adhesion of the textile web to the surface of the printing table. The continuous movement of the conveyor belt results in high uniformity of the adhesive layer on its surface and thorough washing of the belt during printing. The conveyor belt drive of the printing table is controlled by a microprocessor. Each printing group of the machine consists of a device for fixing a flat template over the printing table, a doctor mechanism and a system for distributing printing ink in the templates. The doctor mechanism is equipped with an individual drive from an alternating current electric motor with a reducer, which makes it possible to set a different number of passes of the doctor in individual templates depending on the nature of the pattern. Electronic control of the doctoring mechanism provides soft acceleration and stopping of the double mechanical doctoring. Changing the speed of the rake within the range of 2.4 - 6 m/s is carried out using the control panel of each printing group. The set mode of rake movement can be automatically transferred to all other printing groups. The device for washing the transport belt from printing ink and adhesive composition consists of rotating cylindrical nylon brushes and circulation system of washing solution. For drying the printed fabric, the printing machine is coupled with a nozzle drying machine. By the level of printing automation this model is the most perfect, as it provides control and automatic control of all working bodies of the machine. The machine is available in several modifications with the working width of the printing table 1600, 1900, 2300 and 3200 mm. The length of the machine depends on the number of templates installed on it and varies from 14000 to 46000 mm. The maximum number of templates is 24.

Table 7
Technical data of machines with flat mesh gauges

Indicator	"Meccanotessile" (Italy)	"Stork" (Netherlands)
Maximum print width, mm	1600	1620
Number of passes (colors)	8	8
Number of rakley strokes	1-6	1-7
Ramp size, mm	700-1000	700-1000
Ramp accuracy, mm	± 0,1	± 0,1
Table blade speed, m/min	5-20	4-20
Installed power of electric motors, kW	55	50
Evaporation capacity of drying chamber, kg/h	380	170
Length of fabric filling in the dryer chamber, m	26	9
Maximum temperature in the drying chamber, °C	150	160
Specific consumption per hour:		
- vapor, kg	700-750	600-650
- water, m^3	1.2	1.1
Overall dimensions, mm		
- length	32315	28800
- width	3200	4560
- height	2870	4025

22 Machines with rotary grid patterns

Rotary printing presses of two types are widely used: with horizontal arrangement of the printing table and with printing templates arranged around the central drum.

The main advantage of this type of equipment compared to flatbed printing machines is the printing speed, which reaches 100 m/min.

Figure 43. Rotary mesh templates

The principle of printing on rotary presses:
- the role of the printing roller is played by a hollow seamless metal perforated cylinder with a wall thickness not exceeding 0.2 mm. Inside the templates, a doctor system is installed to force the printing ink through the perforations.

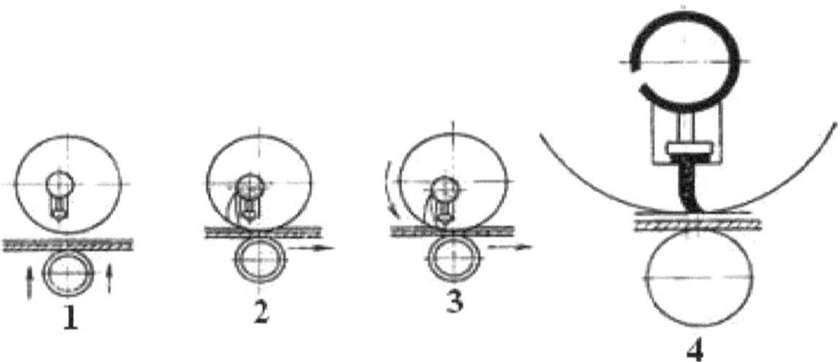

Figure 44. Principle of operation of the doctor and template of a printing press with cylindrical mesh templates

The doctor mechanism can be made in the form of steel or rubber blades located at a certain angle to the template, or in the form of special hollow nozzles for injection of printing ink onto the fabric. Templates are fixed in the working position with the help of template holders located on both sides of the printing table, and depending on the width of the machine, they are equipped with a one-way or two-way drive of printing templates.

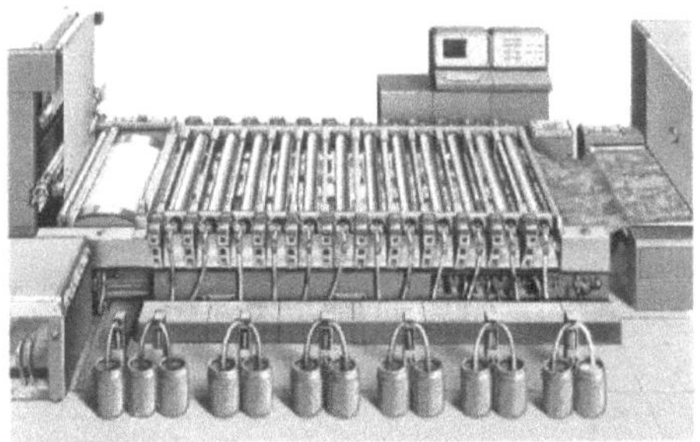

Figure 45. General view of the machine with rotary mesh gauges

The most widespread are printing rotary machines of the Dutch company "Stork", the Italian company "Mecannotessile", the Austrian company "Zimmer" and the Swiss company "Buser".

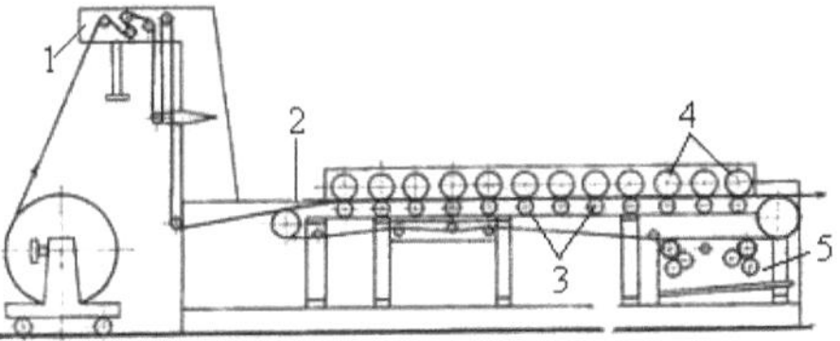

1 - introductory device; 2 - fabric; 3 - printing table; 4 - mesh templates; 5 - device for washing and dewatering of conveyor belt

Figure 46. Diagram of the printing machine with cylindrical mesh templates

Stork printing machines. These machines are characterized by low metal intensity, energy consumption, low labor intensity and duration of the process of changing the pattern; low pressure of the template on the fabric, more economical consumption of dye. In machines of this type cylindrical mesh templates are located above the belt conveyor, which is made of rubberized fabric and acts as a printing table. Inside the template is inserted a doctoring mechanism for feeding and wiping the printing ink. The contact of the templates with the fabric is realized by means of pressure rollers. The forces of pressure of these rollers to the conveyor belt are regulated by devices for each template; this allows to ensure uniform application of ink on the surface of the fabric. Under the lower branch of the belt conveyor there is a unit for washing and cleaning it from traces of glue and paint. The fabric from the roll on the guide rollers goes straight to the conveyor belt and is glued to it with thermoplastic adhesive by means of a special mechanism and a heater. The printed fabric when leaving the machine is easily removed from the surface of the belt conveyor, put on the fabric conveyor and with it goes to the nozzle dryer. The conveyor is made of a polyester mesh fabric, which allows air to pass freely when nozzle blowing. After drying, the fabric is removed from the conveyor and stacked in a cart by a fabric stacker. The printing machines are characterized by increased versatility. They use a computerized control system for all printing processes, including pattern etching. All the production program - information can be recorded in the computer memory, which makes it possible to quickly and accurately reproduce the printing of high-demand patterns and to carry out pattern changes.

Table 8
Technical characteristics of printing machines with rotary templates of the company
"Stork."

Parameter	Indicator
Width of processed fabrics, mm	1620
Linear velocity of main tissue-conducting organs, m/min	4-80
Number of colors	12
Number of webs to be processed	1
Dryer with evaporation capacity, kg/hour	945
Maximum temperature in the drying chamber, °C	150
Filling length in the drying chamber, m	28,2
Installed motor power, kW	101,8
Specific consumption:	
- steam, kg/h	700
- water, m /h^3	1,25
Overall dimensions, mm:	
length	21100
width	6160
height	3645

Righiani printing presses. Righiani manufactures equipment for all known printing methods, namely printing with flat and rotary templates, as well as digital. Rotary printing machines are widely used and have proven themselves.

Table 9
Technical characteristics of the printing machine with rotary templates of the company
"Rigiani."

Parameter	Indicator
Number of templates, pcs.	8-12
Working width, mm	1800
Printed image size, mm	640-1018
Maximum speed, m/min	80
Electricity power consumption, kWh	121,9
Specific consumption	
-steam, kg/h	1200

-water, m^3/h	3
-air, m^3/min	1,2
Evaporation capacity of the dryer, kg/h	825
Overall dimensions, mm:	
length	27650
width	6050
height	4130

23 Transfer printing equipment

For this type of printing equipment is produced periodic and continuous. Of the batch equipment, the most common are various presses consisting of two plates, one of which is metal, heated, and the other has an elastic coating. Of greatest interest are presses with automatic regulation of temperature and pressure between the plates. It should be noted that this is a low-productivity equipment, used almost exclusively for printing piece goods (shirts, ties, scarves) at the enterprises of the haberdashery industry.

The development of continuous thermal calenders contributed to the rapid introduction of transfer printing into the practice of textile printing. Although these calenders differ considerably in design and technical and economic indicators, they all operate on the same principle: the transfer paper, printed with the printed side on the front side of the fabric, is pressed together with it by a pressure belt to a rotating metal drum heated to a certain temperature.

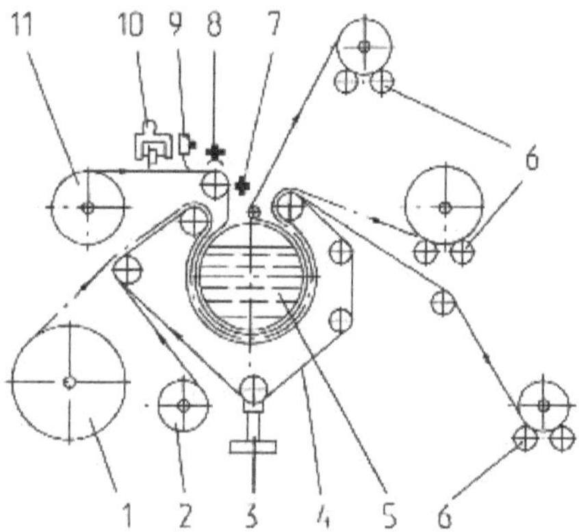

1 - fabric; 2 - paper acting as a layer; 3 - tensioning device; 4 - pressure belt; 5 - oil-heated cylinder; 6 - rolling device; 7,8 - devices for regulation of fabric and paper feeding; 9,10 - knife devices for lateral and cross cutting of paper; 11 - transfer paper.

Figure 47. Schematic diagram of the transfer thermal calender by "Monti" company

To protect the pressure belt from contamination by dye vapors that penetrate during dye sublimation, a protective layer of paper is placed between the fabric and the belt. The temperature of the rotating metal drum is selected depending on the range of fabrics to be processed and the properties of the dyes used. It usually lies within the range of 150-250°C. The printing speed depends on the diameter of the drum and the duration of contact of paper and fabric with the heated surface of the drum. The speed varies smoothly in the range of 3-30 m/min. An endless pressure belt wraps around the metal drum. The transfer paper, tissue and protective paper overlap just before entering the printing area, so there is no need for a mechanism to straighten the paper and tissue. The use of vacuum technology can speed up the transfer process to 25 - 35 m/min and reduce the temperature on the calender by an average of 15-25 °C.

Table 10
Technical characteristics of Stork thermal transfer calenders

Parameters	Types of calenders		
	TS-171.	TS-131.	TS-131.
Maximum transfer width, mm	1800	2000	1800
Diameter of heated cylinder, mm	1800	1350	760
Temperature fluctuations on the cylinder surface, °C	±1	±1	±1
Pattern translation speed, m/min	8-12	18-20	5,9-11,5
Cylinder heating	thermo oil	thermo oil	thermo oil
Overall dimensions, mm:			
length	2500	3000	2000
width	3200	4000	2700
height	2500	2500	1750

24 Inkjet printing machines

The first inkjet machine in the modern sense is the TruColor TCP2500, demonstrated by Stork at the ITMA in Hanover in 1991. It was a four-color machine designed to produce fabric samples according to specified patterns, with a working speed of less than 1 m^2 /h. Inkjet printing has the following advantages: - unlimited number of colors and shades in a pattern;
- unlimited complexity of drawing, up to reproduction of watercolor drawings and photographs;
- almost instantaneous reproduction of the pattern directly on the fabric, which makes it possible to prepare collections of samples for exhibitions or customer approval extremely quickly;
- the possibility of very fast change of patterns and colors;
- virtually waste-free production of small batches (700 - 1000 m);
- excellent reproducibility.

Inkjet printing is held back by three very significant factors:
- low operating speeds, in the order of 50 - 150 m /h;[2]
- labor-intensive cleaning of working heads (nozzles) from paints;
- high cost of ink, and in connection with this - high cost of m^2 fabric.

Reggiani, which produced the first industrial inkjet printing machine, presented the latest development DREAM ("dream") with a working speed of 150 m^2 /h with a width of printed fabric of 1.6 meters. The Italian firm Robustelli released the machine "Monna Lisa" with 8 base colors, working width of 1.6m and working speeds of 26-78 m^2 /h with a resolution of 720 and 360 dpi* respectively, created in cooperation with the Japanese corporation "Epson".

Zimmer's Chromotex was created as an application of modern digital technology to the company's extensive experience in flat pattern and rotary printing. As a result, the amount of ink applied to the fabric is at the level of pattern printing, which contributes to better penetration of ink into the thickness of the fabric, and this is especially important for heavy and dense fabrics for special purposes - auto upholstery, furniture.

The Osiris Isis machine (Holland) is interesting due to its width of 3.2 meters. The Artistri 2020, a joint development of DuPont and Ichinose, attracts with its economic efficiency.

Inkjet printing presses are differentiated:
- by purpose (for making samples, for exclusive printing - silk scarves, ties, as well as industrial purposes);
- operating speed - from 20 to 150 m /h;[2]
- working width - up to 3.5 m;
- resolution up to 780 dpi*;
- type (thermo-, piezo-, combi-) and number (4-24) of working heads;
- number of basic ink colors (4-8);
- software.
*dpi = dots per square inch.

Table 11
Technical characteristics of the industrial digital printing machine of the company Cromotex "Zimmer"

Parameters	Indicator

Maximum fabric width, mm	2250
Maximum pattern width, mm	1850
Fabric thickness, mm	0-45
Printing Modes	4 color heads or 4 color detection channels
Start-up time, min	10
Color change and rinsing time, min	20
Ink type	It is possible to use inks based on cationic, disperse, pigments, acid, active dyes
Applicable programs	Zimmer Windows printable programs
Compatibility	TIFF - format from any input data
Air pressure, bar	8
Water pressure, bar	3
Water consumption during color change, l	50

25 Dye fixation equipment

After printing and drying, the dye is located in the film formed by the printing ink on the surface of the fibers. To facilitate the diffusion of the dye from the film into the fiber, the fabric is treated in a humid environment at high temperature in a special equipment - a ripper. In the ripper, moisture condenses on the cold fabric, which promotes swelling of the fiber. The water sorbed on the fiber dissolves the dye and auxiliary substances included in the printing ink. The process of dissolution of auxiliary substances is accompanied by the release of heat, which contributes to the swelling of the thickener. At the same time, the dye diffuses from the printing ink layer into the micropores of the fiber material and fixes the dye on the fiber.

Fixation of dyes on the fiber after printing is done:
- in a saturated vapor environment;
- of superheated steam;
- of hot air.

The fabric is treated at atmospheric pressure in saturated steam at a temperature of 100-106 °C or in superheated steam at a temperature of 150-190 °C. The duration of treatment depends on the class of dyes used and ranges from 8 to 40 min in the first case and from 3 to 10 min in the second case. For certain classes of dyes and types of printing, hot air with a temperature of 150-210°C can also be used as a fixing medium with a treatment duration of 5-3 min. Heating with IR rays for 8-12 s.

The basic requirements for modern mature equipment are as follows:
-the possibility of fixation in a continuous way of dyes of all classes on various fiber materials;
- variation in a wide range of fixation durations;
- Textile material can be processed with minimum tension or in a free state
- use of saturated and superheated steam as a heat carrier.

At present, the enterprises have installed recovery and universal rippers. Depending on the properties of the printed materials, the rippers of various designs are used, differing in the wiring scheme:
- on rollers - for materials with non-deformable structure;
- curtain loops - for materials with a moving structure.

Restoration rippers. Restoration rippers are designed for continuous processing of dry printed fabrics, for the development of which reducing compositions are used, as well as for those printed with cube dyes and for obtaining an etching on the colored background. The mill is a rectangular chamber assembled of cast-iron plates insulated from the outside. The ceiling plates are hollow inside and are heated by steam. The inner volume of the maker is divided into three unequal parts:
- the steam chamber;
- pre-camera;
- cooling chamber.

The cooling chamber is used to cool the steamed hot fabric before it leaves the kiln. The pre-chamber is designed to capture and remove vapor coming out of the slit through which the fabric is introduced into the ripper. Fabric transportation in the ripper is carried out on two rows of guide rollers. In order to reduce the tension of the fabric web during steaming, every third roller of the upper row is a drive roller. The fabric is steamed in an atmosphere of saturated steam with a humidity of 99.7 - 99.8% at a temperature of 102 - 103°C.

The main disadvantage of rippers with vertical fabric guide is the alternating contact of the fabric with the guide rollers from the back side to the front side. Numerous contacts of the printed side of the fabric with the surface of the guide rollers can smear the pattern and contaminate its background. These disadvantages are eliminated by the spiral fabric feeders. Due to the spiral threading the printed fabric touches the guide rollers only with the wrong side. This eliminates the contamination of the latter with printing ink and the contamination of the background.

Curtain rippers. Curtain rippers are used for processing fabrics made of chemical yarns, natural silk and fabrics with easily deformable structure. The rippers can be single or double web. They can differ in the filling length of the steaming chamber, mainly from 150 to 600 meters. Structurally, burners differ in the system of steam supply, the character of steam medium circulation, the system of loop closing on the roller conveyor. Steam superheating can be carried out by different types of heat carriers by oil, gas or electric heating.

At the enterprises of the silk industry now widely used ripeners firms "Arioli" and "Stork", differing in the design of the steam chamber. In Arioli mills the steaming chamber is made in the form of a box open from below, in Stork mills the steaming chamber is of the closed type.

The curtain ripper of the company "Arioli" (Italy) is designed for processing fabrics in saturated and superheated steam environment. Main parts:
- a filling device, which is a system of rollers, on which the fabric in a straightened state enters the steaming chamber;
- Steam chamber for processing fabric moving in the form of free loops, screwed on stainless steel rods, which are fixed on a conveyor with a DC motor drive. The mechanism for winding loops includes a movable compensator and a gear for moving up and down synchronously with the forward movement of the chain with conveying rods. The steam chamber has double walls and a double lid. Saturated steam is produced by boiling water between the double walls of the chamber, the water level in which is usually 40 - 45 cm. To work with superheated steam, the kiln must be equipped with a steam superheater, in which one of the heat carriers can be used. The superheated steam enters the chamber, from the walls of which there is a preliminary
the water is removed;
- post-processing fabric picking system with rollers and rollers driven by a DC motor, a mechanical feeler gauge to control the chamber filling and automatic increase of the picking speed. The fabric is placed in the cart by means of a self-stacker.

Table 12
Technical characterization of the Vapolitermotex curtain type milling machine v. Arioli.

Parameter	Indicator
Width of rollers, mm	3500
Fabric threading length, m	200
Maximum fabric width, mm	1600 x 2
Number of webs to be processed	2
Hinge length, mm	
maximum	2500
minimal	2250

Speed, m/min	5-50
Processing time, min	4-40
Vapor medium temperature, °C	
saturated steam	102 - 106
superheated steam	165-190
Steam consumption, kg/h	500-700
Installed power of electric motors, including superheater, kW	210
Overall dimensions, mm	
length	12200
width	4500
height	4180

Stork's loop drill. Main parts:

- filling device, which is a frame with width gauges, guides and driven rubberized pulling rollers, the speed of rotation of which is controlled by variators;

- Steam chamber for processing the fabric, which is wrapped with a looping roller of a special stainless steel profile, through which the fabric is fed to the rods moved by a transport chain. The looping is carried out by a rod moving upwards on an inclined guide. The fabric exits the dryer through an outlet device with a Teflon seal. On the bottom of the steaming chamber made of chromed steel, two circulation fans are installed, which circulate steam inside the chamber of the burner, providing together with steam-water injectors to maintain constant conditions of fixation of the treated fabric. In addition to recirculation and reconditioning, a system of counterflow of the working medium and fabric is provided in the burner;

- The selective device ensures stacking of the fabric in the cart. A special fabric loading mechanism synchronizes fabric feeding and picking.

Table 13

Technical characteristics of Stork loop cutters

Parameters	Types of rippers	
	HSIII/3600/215	HSIII/2800/150
Width of rollers, mm	3600	2800
Length of fabric to be tucked in, m	215	215
Maximum width of fabrics to be processed, mm	1800 x 2	1200 x 2
Number of webs to be processed	2	2
Hinge length, mm		
maximum	3000	3000
minimal	2600	2600
Speed, m/min	5-50	5-50
Processing time, min	4,3-40	4,3-40
Vapor medium temperature, °C		
saturated steam	101-103	101-103
superheated steam	up to 185	up to 185
Steam consumption, kg/h	700	550
Installed power of electric motors, including superheater, kW	210	200

Overall dimensions, mm		
length	16934	15592
width	5492	4880
height	5194	5194

Universal curtain-type ripper ZZU-4/260. It allows processing of fabrics in the environment of saturated and superheated steam, as well as in the environment of azeotropic mixture of saturated or superheated vapors of benzyl alcohol and water.

The fabric is introduced into the ripper chamber by means of the filling rollers, fabric guide and pulling drum. The fabric is then looped on the rollers, guided through the chamber, exited through the cooling drum and placed in the cart. A humid steam environment is created by heating the water mirror with blind steam, as well as humidifiers supplying sharp steam. To obtain superheated steam, an electric steam superheater built into the air removal chamber is used. The ceiling of the kiln is heated with steam. The temperature of saturated steam is 102-106 °C, superheated steam - up to 180 °C. The length of the chamber is 180 m, the speed of fabric movement is 8 - 80 m/min. Overall dimensions of the chamber are 18800×4900×4200 mm. Installed power of electric motors 16.7 kW, electric heaters - 240 kW.

Questions for knowledge control

1. What are the commonalities and differences between the dyeing and printing processes? How is this taken into account when designing a fabric patterning process?
2. What substances are used as thickeners for printing inks? Characterize the requirements for them. Can starch-based thickeners be used when printing with active dyes and pigments?
3. List the main methods of patterning on fabric. Characterize the area of their application, advantages and disadvantages.
4. Does the nature of the fiber material from which the fabric is made affect the choice of printing method? Which method of patterning is preferred for cotton fabrics, fabrics made of protein fibers, knitted fabrics, fabrics made of thermoplastic chemical fibers?
5. What is the purpose of processing fabrics in rippers after they have been printed? What types of rippers are used in finishing production, which of them are used for processing of fabrics printed with active, dispersed, cubic dyes, cubosols, pigments, black aniline?
6. Is the washing technology of fabrics printed with different nature of dyes the same? How do washing conditions and quality affect coloristic, performance and hygienic properties?

26 Printing knitted fabrics

The range of printed knitwear includes linen knitwear, women's dresses, blouses, sportswear and hosiery. The physical and chemical processes occurring on the fiber during printing are identical to dyeing. The same dyes are used for patterning, the mechanism of fixation on the fiber is preserved. The main differences are in the method of applying the dye to the surface of the textile material and in the hardware design of the process. Drawing a pattern on knitted fabric is associated with certain difficulties due to the structural features of knitted fabric: high stretchability, the presence of a strongly twisted selvedge or lack thereof, the relief of the surface. The consequence of high stretchability is often a distortion of the shape of the pattern, especially when printing fabrics of lightweight structures. Printing difficulties are also caused by a strongly curling edge. This prevents the printing device (print roller or template) from adhering tightly to the web across its entire width. For this reason, printing presses must be equipped with devices that allow the selvedge to be straightened. Good results can be achieved by gluing the edges.

Another feature of the printing process in comparison with dyeing is the need to obtain clear contours of the pattern, as the printing process is considered as local dyeing of individual areas of the knitted product. Thickened dyeing solutions - printing inks - are used for drawing the pattern. Printing inks are compositions containing dye, solvent (usually water), substances that create conditions for the fullest possible fixation of the dye on the fiber, and thickener.

27 Preparation of knitted fabrics for printing

The semi-finished products subjected to printing must have good capillarity so that the dye can easily penetrate into the fiber from the printing ink, as well as smooth appearance, stable dimensions, absence of skewness of the loop rows to avoid distortion of the shape of the pattern. These properties are imparted during the preparation of semi-finished products for printing. Preparation includes boiling, bleaching, drying-splitting of basic knitted fabrics or drying and calendering of circular knitted semi-finished products. The latter can be printed in cut and uncut form (i.e. as a tube). When applying the pattern to the uncut fabric, a number of problems arise: the fabric can not be glued to the surface of the table, as the reverse side is also printed, there is a need to ensure the coincidence of the pattern on one and the other side of the tube and, especially, on the fold; possible deformation of the fabric due to its uneven shrinkage; reduces the productivity of the printing machine, because the fabric is passed through it twice. Therefore, in most cases, webs are printed in cut form. Cutting is one of the preparatory operations and is carried out on special machines, where, in addition to cutting, the web is straightened and, if necessary, the edges are glued. Some types of semi-finished products, depending on the purpose and nature of the fiber can be subjected to additional treatments. For example, wool fabrics are chlorinated before printing, and lavsan fabrics are thermofixed.

28 Methods of patterning on knitted fabrics

In the knitting industry all known methods of printing ink on textile material can be realized: hand printing, airbrush method, polychromatic dyeing, printing with metal engraved rollers, mesh templates, transfer thermal printing method. Machines with metal engraved rollers, which have high productivity, have not found widespread use in the knitwear industry due to the large deformation of fabrics passing through the machine under strong tension. Manual printing with the help of mesh templates is sometimes used for piece goods (hosiery, sportswear, top knitwear). The most widespread application in knitwear production is the method of photofilm printing (with mesh templates). For this purpose two types of machines are used: with flat and with rotary mesh templates.

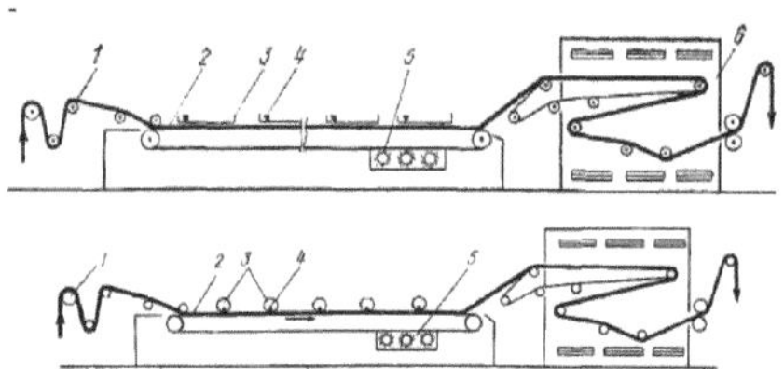

Figure 48. Printing presses: with flat and cylindrical mesh templates

Common for both types of machines is the presence of a printing table, the role of which is performed by an endless conveyor belt 2. After passing the filling device 1, the web is glued to it with a special adhesive and is transported through the working area of the machine with its help. Then at the exit from the printing zone the web is removed from the surface of the printing table and enters the drying chamber 6. The surface of the conveyor belt contaminated with printing ink is washed in the bale washer 5, dried and returned to the beginning of the table.

The machines differ in the design of templates 3 and doctoring mechanisms 4, wiping the printing ink through the sieves of templates. The template is a rectangular frame on which a kapron or polyester mesh is stretched. The size of the frame of the template is determined by the width of the machine, and the length should be a multiple of the pattern. Two or three layers of varnish, which is a solution of perchlorovinyl resin in butyl acetate, are applied to the grid of the template. Then the template is covered with photosensitive emulsion, dried and installed on a photocopying machine. Copying is made from rapportment cribs. On them the areas of the pattern, which should not be exposed to light during exposure (illumination by a metal hologen lamp), are covered with black ink. When light hits the unprotected areas of the photosensitive layer, it blots out to a water-insoluble compound. No blowing occurs in the protected areas, and the photosensitive emulsion in these areas is washed away when rinsed with water. The varnish from a grid of a template is removed by a swab moistened in butyl acetate. The number of templates necessary for reproduction of a drawing is determined by the number of colors in it. Templates are installed above the conveyor belt and at the moment of its stop automatically lowered onto the canvas. At this point, turn on the mechanisms of movement of the rakel, which, having made one or two passes, wipe the printing ink through the sieve template. At the end of the rakli templates are automatically raised, and the web together with the conveyor belt is moved to the size of the rapport (the length of the template). One template forms an imprint of a pattern detail of one color, a series of templates reproduces the pattern as a whole.

The main disadvantages of these machines are the long length (25-30 m) and low productivity, which is due to the time spent on stopping, lowering the templates and operation of the doctoring mechanism. This led to the creation of machines with rotary (cylindrical) mesh templates. The basis of the cylindrical template is a seamless hollow cylinder made of chrome-nickel mesh. The surface of the cylinder is coated with a layer of photosensitive emulsion and dried. Then a positive film with the pattern is placed on the nickel sleeve and exposed. The sleeve is placed in a bath with water (t=30° C) for 5-6 minutes for lacquer swelling and a strong jet of water is used to wash it off from the places covered by the drawing and not exposed. The final operation is heat treatment of the sleeve at 180° C for 1 hour in order to polymerize the lacquer film in the places of the pattern and to give it the necessary chemical and mechanical strength. End rings are glued into nickel sleeves of templates, which are installed in the template holders of the printing machine. Printing ink through hoses with the help of pumps is fed inside the templates and wiped with the help of a fixed fixed inside the rotating template rakli 4. The number of templates corresponds to the number of colors. Machines with cylindrical templates work with higher speed, have smaller dimensions, can print patterns of almost all types on knitted fabrics of any assortment.

29 Web handling after printing

Drying the fabrics after printing is necessary to prevent smudging of the pattern and to preserve its contours in subsequent processing. In the drying machine, the fabric is blown with hot air (90-130° C). The material must not be pulled or deformed. After drying, the dye is in a thin film of dry printing ink on the fiber surface. In order to ensure the conditions for a firm fixation of the dye, a heat treatment is necessary, during which the dye must transfer from the printing ink layer to the fiber. The heat treatment conditions depend on the type of fiber and the class of dyes used for printing. Fixation of the majority of dyes on natural fibers is carried out in saturated steam at 100-105° C or superheated steam at 140-170° C. In the process of steaming under the action of moisture and heat the film of thickener swells and conditions are created for diffusion of dyes first inside the layer of printing dye to the surface of fiber, and then sorption and diffusion inside the textile material. In parallel with this, a number of chemical transformations take place: reduction, oxidation, dissolution, formation of high molecular weight resin. The nature of these reactions is determined by the mechanism of fixation of a given class of dye. In some cases, dye fixation is carried out by dry hot air at 150-200° C (heat treatment). At knitting enterprises for steaming of fabrics the equipment of periodic (boiler-type apparatuses) and continuous (loop rippers) action is used.

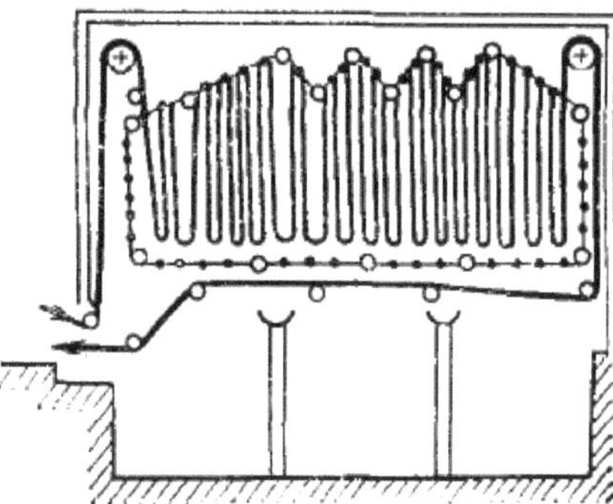

Figure 49. Loop borer

The burner is a rectangular chamber isolated from the environment. It is supplied with saturated or superheated steam, in some cases with hot air. To prevent dripping, the walls and ceiling of the chamber are made of heated plates.

The fabric is inserted through a narrow slit into the ripper and, with the help of a curtain device, is hung on rollers attached to endless conveying chains, forming a free loop. This avoids pulling and warping, which is particularly important for knitted fabrics with lightweight structures and textured yarns, which are prone to warping. After steaming or heat treatment, the knitted fabrics are washed to remove the thickener, unfixed dye and TVB included in the printing ink. The optimum washing method is to process the unfolded fabric in a continuous process. For this purpose the same lines are used as for washing-relaxation: MP-220-T, machines of foreign companies. Washing conditions depend on the class of dyes used for printing.

30 Printing of piece goods

There are mainly two methods used for printing piece goods:
- direct printing using mesh templates;
- transfer printing.

For direct printing, machines with flat mesh templates are used or printing is done manually with mesh templates on tables. Parts are printed after the product has been cut. A grid pattern printing machine has 10 small flat printing tables mounted on an endless chain conveyor that moves them along with the parts through all working positions. The part is placed on the table, which is pre-lubricated with adhesive. This is necessary to prevent the part from shifting during printing and transportation, as well as to prevent it from coming off the table plane and sticking to the template during the printing process. Next, the part on the stage is moved to the first template, the template is lowered, the rakla makes a specified number of passes, the template is raised and the part is moved to another template. At the end of the complete printing cycle, the parts are removed from the tables and placed on the conveyor line of the drying machine, installed next to the printing machine. After printing, the cut parts are transferred to the sewing shop, where they are used to make products, which then undergo final finishing.

In accordance with fashion requirements, printing of hosiery has been introduced in recent years. They are printed manually using mesh templates. The product is put on a mold and printed using pigment dyes that do not require washing. Drawing can also be carried out by the method of transfer thermal printing. The essence of transfer printing is that the pattern is applied to the backing paper, and then from the paper is transferred to the textile material. Two types of transfer printing are used in knitwear production. The first one is based on the use of thermoplastic polymers. In this case, the pattern is printed on the backing paper with printing ink containing dye, binder and thermoplastic polymer. The latter in the process of transfer under the action of high temperatures softens, is transferred from the paper to the textile material and glued to it, and the binder provides a strong fixation of the thermoplastic film containing dye. To transfer the pattern from paper to fiber use presses with the upper heated metal plate, the temperature of which should not be lower than $210°$ C. The duration of pressing is 20-30 seconds. This method of thermal transfer can be used for printing products made of both natural and synthetic fibers. The second method of transfer printing is based on the use of sublimating dispersed dyes, so it is called "sublistatic". The widespread use of this method is due to the rapid growth in the production of synthetic fibers, especially polyester, printing of which by the traditional method causes great difficulties. The method of "sublistatic" is based on the ability of some dyes when heated (to a temperature of $180-200°$ C) to sublimate - to change from a solid state to a gaseous one - and in the form of vapors to be sorbed on the textile material, to penetrate inside the fiber and to be fixed on it. For coloring knitted fabrics and piece goods this method of printing is of great interest, because it allows to exclude the operations of steaming and washing. Method "sublistatic" with great accuracy reproduces complex patterns, provides high quality when printing primer patterns, allows you to get a clear contours (even thin lines) and quickly change patterns. Transfer of the pattern from paper to canvas is carried out on special equipment - calendars. Two types of calendars have become widespread:
- Thermal calenders with heated metal roll and cloth;
- vacuum calenders.

Diagram of the calender with cloth:

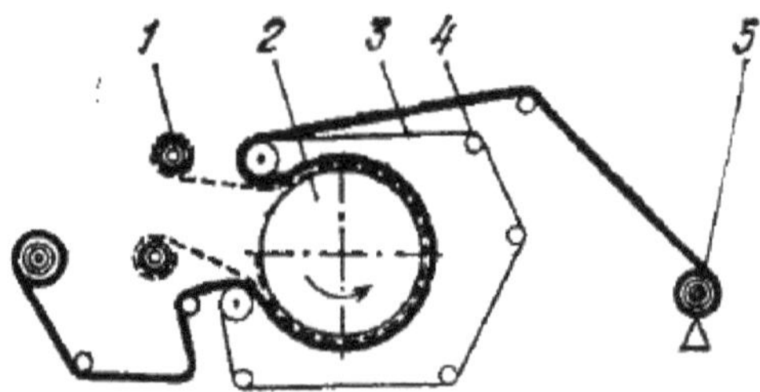

Figure 50. Calender for thermal printing with felts

The main working body of this calender is a large metal heated cylinder 2, which acts as a table for applying the printing pattern. Around the cylinder runs an endless heat-resistant cloth 3, moving on guide rollers 4. The paper with the pattern 1 and the cloth 5 are tucked between the hot roll and the cloth, which tightly presses the paper to the cloth and to the heated surface of the cylinder. The temperature of the roll surface is regulated in the range of 150-230° C, the duration of contact is determined by the speed of rotation of the cylinder and is 15-45 seconds.

The disadvantage of such a calender is the change of the knitted fabrics' grift due to the strong pressing of the fabric by the cloth to the heated surface of the cylinder. In addition, as a result of uneven pressing, there may be uneven transfer of the pattern from the paper to the fabric. From this point of view, vacuum calenders operating on the principle of vacuumization are more promising.

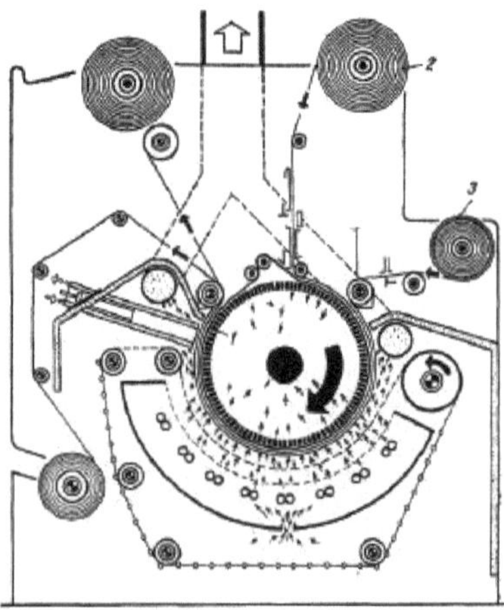

Figure 51. Vacuum calender for thermal transfer printing

A vacuum of 8-13 kPa is created inside the perforated cylinder 1. Heating is carried out by infrared radiators installed outside. Paper 3 and cloth 2 go around the perforated cylinder and move with it, pressing against its surface due to the pressure difference between outside and inside it. The use of the vacuum principle accelerates the sublimation of the dye and facilitates its transfer from the surface of the paper to the fiber. As a result, dyeing is improved, color brightness and saturation are increased, and strength properties are improved. The absence of strong pressure, lower temperature and shorter contact time ensure a soft grift and preserve the bulkiness of the web. Presses with an upper electrically heated metal platen are used to transfer the pattern from paper to piece goods. Both cut parts and finished products can be printed. Thermal printing is subjected to fabrics and products made of almost all synthetic and triacetate yarns, as well as from a mixture of synthetic fibers with natural fibers, if the content of the latter does not exceed 25-30%. The most widespread is thermal printing of knitted fabrics and products made of polyester fibers. The ability of dyes to sublimation depends on the structure of dye molecules, their molecular weight and polarity. There are 40 brands of disperse dyes, 2 acid dyes, one cube dye and 14 cationic dyes recommended for printing applications. In addition to the dye, the composition of printing ink applied to paper contains a solvent (water, organic solvent or a mixture of both) and a thickener. Special requirements are imposed on the thickener for transfer printing: it should be heat-resistant, it should not interact with the dye, the thickener film should not prevent dye sublimation. An important factor affecting the quality of transfer printing is the quality of the backing paper. It should be heat-resistant, have a smooth surface, evenly wetted by the printing compound, remain on the surface of the paper without penetrating inside it. Special low-porosity paper with a glossy surface is produced for thermal printing. Drawing on paper can be applied by methods used in the printing industry (gravure printing, offset and flexographic), or on machines with grid templates used in traditional printing. Improvement of thermal transfer printing is going in the direction of improvement of dye release forms, development of new types of reusable substrates based on heat-resistant polyamides, siliconized paper, development of transfer printing methods for products made of natural fibers.

31 Types of printing

According to the nature of the pattern and the area occupied by it on the fabric, a distinction is made between white-ground, semi-ground and ground printing. In knitwear production, the most widespread is white-ground printing, in which the area of the fabric occupied by the pattern does not exceed 20-30%. It is more often used for linen products. Ground patterns (more than 60% of the area) and semi-ground (40-60%) are used relatively rarely (excluding transfer thermal printing), which is due to the change in the fabric grain and the difficulty of obtaining uniform printing due to the peculiarities of the structure of knitted fabrics. However, for top knitted garments they are of great interest. Printing can be done on bleached or dyed fabrics. Accordingly, a distinction is made between direct, etched and reserve printing.

In *direct printing, the printing* ink is applied to a bleached or light-colored canvas. In the latter case, the printing ink must be a richer tone or other color, opaque, and the background color must not blend with the color of the printing ink.

In etching *printing, the* pattern is applied to a pre-dyed knitted fabric or product with a printing ink containing substances capable of destroying the background coloring. For pre-dyeing use dyes capable of etching (direct, insoluble azo dyes, disperse and acid azo dyes). The peculiarity of these dyes is the ability of their chromoform azo groups to break down under the action of reducing agents:

$$R_1-N=N-R_2 \xrightarrow[\text{восстановление}]{2H_2} R_1-NH_2 + R_2-NH_2 .$$

As a result, colorless amines are formed, which are relatively easy to remove from the fiber when washing. Etching printing can be white or colored (white or colored pattern is produced on the dyed fabric, respectively). The printing ink contains a reducing agent (rhongalite), an alkaline agent (potash), a thickener and water. Sometimes hydrotropic agents and reduction catalysts are added to the printing ink to speed up the reduction process. Printing ink for color etching printing additionally contains a dye that is resistant to the action of reducing agents or requires reduction for firm fixation on the fiber. The most widely used for color etching are cube dyes. The technological process of etching printing is carried out according to the scheme: put on the cloth or product drawing, dry it, subjected to steaming at a temperature of 100-120° C for 5-15 minutes and washed. The most important operation is steaming, as it is under the action of moisture and heat occurs decolorization (recovery) of the dye. In color etching, simultaneously with recovery, the process of transition of the cube dye into a soluble form and its fixation on the fiber is carried out. In *reserve printing,* printing ink is applied to the bleached fabric and then subjected to dyeing. The printing ink must contain substances that prevent the dye from fixing during dyeing. This is the oldest method of printing, which is the basis of the batik painting technique. A pattern is printed on the product with a substance impervious to dye, such as heated wax, and after the wax solidifies, the product is dyed. In the places where the wax is applied, the dye cannot penetrate into the fiber and fix itself there, so after removing the reserving substance, these places remain undyed. Currently, various chemical substances are used for reserving. Direct printing is the most widely used in the knitting industry, as the etching and reserving methods are quite complicated. However, these printing methods are very interesting in terms of obtaining highly artistic coloristic effects on the details of knitted products.

32 Printing technology of products made of different types of fibers

For printing knitted fabrics made of cellulose fibers (cotton, viscose) it is possible to use cube, active, direct, azoic dyes and pigments. Cubic, active dyes and pigments are widely used in practice. In knitwear production apply mainly rongalitno-potashny method of printing *with cube dyes*. At its realization the printing ink contains a dye (in the form of a paste for printing), a reducing agent (rongalite), an alkaline agent (potash) and a thickener. Paste cube dyes contain special substances that ensure their stability during storage and promote rapid recovery of the dye (antifreezes, antiseptics, catalysts, dispersants, hydrothoric substances). After printing, the cloth is dried, steamed in a reducing burner at 102-105° C for 10-15 minutes and washed with cold water, oxidizer solution to transfer the dye into the original insoluble form, boiling surfactant solution and water.

With active dyes, preferably X-marked dyes, knitted fabrics are printed by a one-step process. The printing ink contains dye, urea, sodium bicarbonate, ludigol, water and thickener. Urea is introduced for better dissolution of the dye, sodium bicarbonate - to create an alkaline environment for the formation of covalent bonds, ludigol prevents the destruction of the dye in the conditions of steaming. Starch should not be used as a thickener, as it forms a poorly soluble compound with the dye, which is difficult to wash off the fiber and gives it stiffness. The best thickener is sodium alginate. The dye fixation is carried out in a steam or thermal burner. When using for fixation of hot air at 150-180° C the content of urea in the printing ink is increased to 150-200 g / kg. The cloth is washed with cold water, a solution containing soap and soda at 95° C and again with warm and cold water. Active dyes allow to obtain colors with high resistance to wet processing, which is very important when printing sports knitwear, such as swimwear. The printing ink contains dye, sodium bicarbonate, urea, ludigol, thickener and water.

Dispersed dyes. Dispersed dyes are most commonly used for printing acetate fiber products. Printing ink contains dispersed dye, dispersant, urea, ludigol, thickener and water. Dye fixation is carried out in saturated water vapor at 102-105° C (20-30 minutes) or superheated vapor at 110-130° C (7-20 minutes). Fixing of the dye on the triacetate fiber products is carried out in thermal ripeners, followed by washing. Knitted fabrics made of *polyamide fibers are* most often printed with dispersed, active and acidic metal-containing dyes. The composition of the printing ink for printing with disperse dyes includes dye, urea, ludigol, sometimes intensifier, water and thickener. *Acidic metal-containing dyes are* more often used in printing knitted fabrics from complex and textured yarns elastic. The composition of printing ink includes dye, intensifier, urea, ammonium sulfate or acetic acid, water and thickener. After printing and drying the cloths are steamed in saturated steam at a temperature of 102-104° C for 20-25 minutes and washed. In recent years, the range of lightweight knitted fabrics made of polyacrylonitrile fiber, designed for sewing dresses, has expanded. Dispersed and cationic dyes can be used for printing such fabrics, but cationic dyes are preferred. The composition of the printing ink includes cationic dye, thiodiglycol, acetic acid, water and thickener. After printing and drying, the fabric is steamed for 20 minutes and washed. When printing knitted fabrics from polyester fiber there are difficulties in achieving the desired degree of dye fixation due to the high hydrophobicity and crystallinity of the polymer. Therefore, the method of transfer thermal printing is mainly used for patterning. When using traditional printing methods, dispersed polyester dyes and sometimes pigments are used to apply patterns to knitted polyester fabrics. The printed ink contains dispersed dye, dispersant, urea, acetic acid, water and thickener. Fixation is carried out in the environment of superheated steam at a temperature of 150-170° C for 6-8 minutes or in the environment of hot air at 180-200° C. In recent years, the range of knitted fabrics made of a *mixture of fibers* (cellulose and polyester, viscose mixed with acetate and polyamide, a mixture of acetate and polyamide) is constantly expanding. In this case, the most convenient option is the use of dyes of the same class, allowing to obtain a uniform color regardless of the chemical nature of the fiber. Pigment dyes have such properties, which are of particular interest for printing knitted fabrics and products made of a mixture of fibers.

33 Printing with pigments

Pigments are dyed, water-insoluble substances that have no affinity for the fiber. They are therefore fixed with polymer film-forming substances, i.e. as if glued to the surface of the textile material.

Both organic and inorganic substances are used as pigments. Inorganic compounds include insoluble metal oxides and salts, microcrystalline soot, metal powders (bronze, aluminum), mica. With their help, unusual coloristic effects can be obtained on the products - printing imitating gold, silver, mother-of-pearl. Organic pigments are organic dyes insoluble in water: azo pigments, polycyclic compounds, phthalocyanine, fluorescent pigments. Pigment dyes can be used to print any kind of textile materials regardless of the chemical nature of the fiber, so they are of particular interest when applying patterns to knitted fabrics and products made of a mixture of fibers. The main factors determining the quality of pigment printing are:
- degree of pigment dispersibility and dispersion homogeneity;
- quality of the binder that forms a polymer film on the fiber;
- the quality of the printing ink thickener.

Coloristic properties depend on the degree of pigment dispersibility and its homogeneity: brightness of color, covering power, resistance of colors to friction. The optimum degree of pigment dispersity is 0.6-2 microns. The quality of pigment printing depends to a large extent on the properties of binders that provide a strong fixation of the dye on the textile material. The polymer film, which they form on the fiber surface, should be transparent, soft and elastic, so as not to give rigidity and not to change the elasticity of the knitted fabric. The polymer should have high adhesion to the fiber, firmly retain the pigment, and have a sufficiently high resistance to physical and chemical influences. It is very difficult to find individual substances that meet all these requirements. Therefore, a mixture of substances including film-forming and crosslinking (mesh-forming) compounds is used as binders for pigment printing. Film-forming substances evenly distribute finely dispersed pigment particles and form a transparent colored film on the fiber surface with adhesive ability. They are used as various thermoplastic polymers (latexes), for example, copolymers of butanediene with styrene (latex SKS-65 GP), acrylic acid esters, etc. Cross-linking (mesh-forming) compounds at heat treatment are able to form a spatial mesh structure, providing a strong bond between the film and the fiber material. Precondensates of thermosetting resins, which molecules contain several reactive functional groups, are used as such substances. Some of these groups interact with the film-forming polymer, others with the fiber, which provides a strong fixation of the pigment. Metazine is most widely used as a mesh-forming polymer. Scientific research aimed at improving the process of pigment printing has led to the creation of preparations that combine the properties of both film- and mesh-forming substances, such as methyl derivative of methacrylic acid amide:

$$CH_2=\overset{\overset{\displaystyle CH_3}{|}}{C}-\underset{\underset{\displaystyle O}{||}}{C}-NHCH_2OH$$

Upon polymerization of this compound, a thermoplastic film is formed, and due to the reactive N-methylene group ($-NHCH_2 OH$) it is firmly fixed on the fiber. Special requirements in pigment printing are placed on thickeners. The use of traditional thickeners leads to the deterioration of the web grift, increases its stiffness, as the thickener is not washed off during washing, since it is insoluble in water. One of the ways to solve this problem is the use of thickeners, the components of which are volatilized from the fiber during drying. Such thickeners include emulsion thickeners, which are viscous systems based on water and insoluble organic solvents. They have all the necessary properties: good thickening ability, achieving clear contours, deeper penetration of printing ink, and most importantly, they do not require washing. However, the high content of solvent (white spirit) in the thickener makes it explosive and pollutes the atmosphere. The solution to this problem was achieved by using synthetic thickeners based on acrylates, produced in our country under the trade names "emukrilov". The formation of polymer thermoplastic film and its fixation on the fiber occurs only at high temperature and in the presence of catalysts (ammonium or magnesium salts). Taking into account the said technological process of printing with pigments is carried out according to the following scheme. The knitted fabric or product is printed with an ink containing pigment dye in paste, thermoplastic polymer, mesh-forming polymer (precondensate of thermosetting resin), catalyst, softener and thickener. After printing, the textile material is dried and heat treated at 140-150° C for 5 minutes. Washing is eliminated when printing with pigments, which is a special advantage when printing piece goods. The drawing can be applied to both white and painted canvas without etching the coloring of the main background. The main disadvantage of pigment dyes is the low resistance of coloring to friction. This is due to the fact that the pigment is located on the surface of the fiber in a thin film of polymer, which is not highly resistant to mechanical effects. Another disadvantage is increased stiffness in the places of patterning, which requires the introduction of a softener and the correct choice of binder and thickener.

34 Direct printing technology in the laboratory

Laboratory work №1. Printing of fabrics with active dyes

<u>Materials and chemicals required:</u>
Cotton cloth 10x10 cm - 6 pcs.
Silk fabric 10x10 cm - 6 pcs.
Woolen cloth 10x10 cm - 6 pieces.
Fabric made of polyamide fibers 10x10 cm - 6 pcs.
Mesh template - 1 pc
Rubberized doctor blade - 1 pc
Drying cabinet - 1 pc
Laboratory drill press - 1 pc
Sodium hydrogen carbonate ($NaHCO_3$) - 50 g
Potassium carbonate - 20 g
Sodium carbonate - 20 g
Sodium chloride - 50 g
Sodium hydroxide 30% - 50 ml
Soda ($Na_2 CO_3$ - 100 g
Active coloring agent - 10 g
Active dye with mark T - 10 g
Urea - 100 g
Ammonium sulphate ($NH)_{42} SO_4$ - 50 g
Alginate thickener - 400 g
Sodium acetate - 50 g
surfactant - 50 g

Active dyes produce colors of high brightness and strength, so they are widely used in the printing process. These dyes are universal and can be used for patterning on fabrics made of cellulose and protein fibers. The main challenge in the development of textile coloring technology is to reduce the proportion of formation of hydrolyzed dye with thickener. For this reason, starch thickener cannot be used for printing with active dyes, because starch like cellulose enters into chemical interaction with the dye, forming a water-insoluble compound that is firmly retained on the fiber. Sodium alginate or cellulose ethers can be used as thickeners. One-stage and two-stage technologies are used for printing with active dyes.

Study of the technology of single-stage printing of fabrics. The different modes of single-stage printing differ in the way the dye is fixed on the textile material:

Figure 52. Schematic of single stage printing

Assignment 1: Study the technology of printing cellulose fiber fabrics in a steamy way.

Progress of the work:
Prepare 50 g of printing ink containing:
Dye - 0.5 g
Water - 5.0 ml
Urea - 3.0 g
Sodium hydrogen carbonate (NaNCO$_3$) - 1.0 g
Alginate thickener - up to 50 g
Procedure for preparation of printing ink: dissolve urea in water, heat the solution to 50-60° C and pour with stirring into a beaker with dye, achieving its complete dissolution (test on filter paper). Immediately before use, add the required amount of sodium bicarbonate dissolved in a small volume of water to the printing ink. Mix the composition thoroughly and apply the pattern on cotton fabric with a grid template.
After printing, the sample should be dried in a desiccator and steamed for 10 minutes in a laboratory steam flask. Next, the printed sample should be washed more carefully than the dyed sample to prevent the white background from being stained by the unfixed dye. For this purpose, the unfolded specimen is first washed with cold and warm water and then boiled in a surfactant solution. Finally, the sample is rinsed again with cold water and dried.

Assignment 2: Study the technology of printing fabrics with active dyes by one-stage thermal process.

Printing ink for printing by thermofixation method should contain an increased amount of urea, which, absorbing moisture from the air, creates on the fabric under conditions of thermofixation the necessary humidity for plasticization of the thickener and penetration of the dye into the pores of the fiber.

Progress of work

Prepare printing ink according to the following recipe, g/50 g:

Active colorant - 0.5

Urea - 15.0

Water - 5.0

Sodium hydrogen carbonate ($NaNCO_3$) - 1.0

Alginate thickener - up to 50 g

The procedure for preparing the printing ink is the same as in task 1.

After printing and drying, the cotton fabric sample is kept in a heat chamber at 160-170° C for 3 minutes, then washed as described in task 1.

Study of the technology of printing fabrics by two-stage method. Two-stage printing method provides separate application of dye and alkaline reagent to fabric. The printing composition has a neutral reaction, which reduces the degree of dye hydrolysis and its interaction with the thickener. Scheme of two-stage printing:

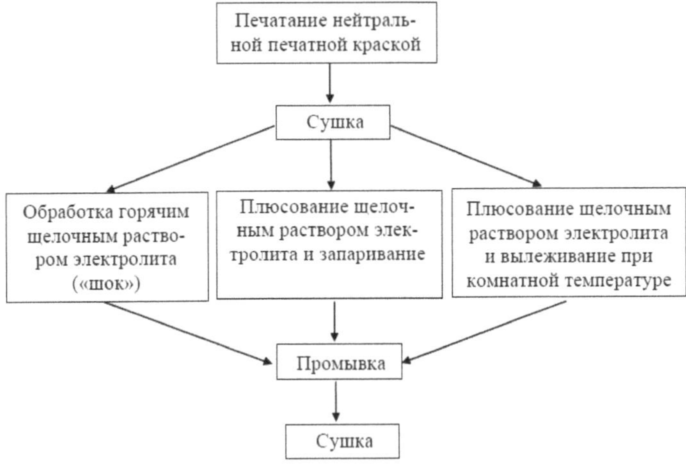

Figure 53. Diagram of two-stage printing

Assignment 1: Study the technology of printing fabrics with active dyes by a two-stage "shock" method of fixation.

Progress of Work:

Prepare neutral printing ink according to the recipe, g/50 g:

Active colorant - 1.0

Urea - 2.0

Water - 5.0

Alginate thickener - up to 50 g

Print the sample of cotton fabric with this composition and dry it. Then prepare an alkaline electrolyte solution (developer solution) of the following composition, g/100 ml:
Potassium carbonate - 5.0
Sodium carbonate - 10.0
Sodium chloride - 10.0
Sodium hydroxide 30% - 3.0
Water - up to 100 ml
The printed sample is treated with developer solution at 95-100° C for 5-10 s, then rinsed and dried.

**Assignment 2: Study the technology of printing of products made of protein fibers (wool
and natural silk) with active dyes.**

In this case, good results are achieved with T-marked active dyes.
Progress of the work:
Prepare printing ink of the following composition, g/50 g:
Active colorant - 1.0
Urea - 0.5
Water - 5.0
Sodium acetate - 2.0
Alginate thickener - up to 50 g
Print a sample of silk or wool fabric with the prepared composition, dry it, steam it in saturated water vapor for 15-20 minutes and wash it. Wash first with cold water, then with warm water (30-40° C) with the addition of surfactant (1 g/l), finally with cold water and dry.

**Assignment 3. Study the technology of printing fabrics and knitwear from polyamide fabrics
fibers.**

The printing technology in this case is close to the technology of printing these materials with acid dyes.
Progress of the work:
Prepare printing ink according to the following recipe, g/50 g:
Active colorant - 1.0
Urea - 2.0
Water - 5.0
Ammonium sulfate $(NH)_{42} SO_4$ - 2.0
Alginate thickener - up to 50 g
Print the sample, dry, steam for 30 minutes, treat in a solution of baking soda ($Na_2 CO_3$, 1 g/L) for 10 minutes, rinse and dry.

Laboratory work No. 2

Printing of cellulose fiber fabrics with cube dyes

Materials and chemicals required:
Cotton cloth 10x10 cm - 6 pcs.

Mesh template - 1 pc
Rubberized doctor blade - 1 pc
Drying cabinet - 1 pc
Laboratory drill press - 1 pc
Cube dye (paste) - 10 g
Cube dye with mark D - 10 g
Glycerin - 20 ml
Potassium carbonate (potash) 1:1 with water - 100 ml
Rongalite 1:1 with thickening - 100 g
Thickener - 500 g
Sodium sulfite 1:1 with water - 100 ml
Urea - 100 g
Catalyst (dimethylglyoxime) - 25 g
Sodium dithionite (hydrosulfite) - 200 g
30% sodium hydroxide - 200 g
Sodium sulfate - 200 g
Potassium bichromate solution 30% - 200 ml
Hydrogen peroxide solution 30% - 100 ml
surfactant - 50 g

Cubic dyes are the most valuable in the technology of printing cellulose fiber fabrics. The resulting patterns are characterized by brightness and durability of colorations. The dyes are represented by a full range of colors and, very importantly, do not paint over the white background. There are several ways of printing with cube dyes, each of which has its advantages and disadvantages: hydrosulfite-rongalite, rongalite-potash and two-stage. The technology of printing by these methods has its own specific features.

Assignment 1. Study the technology of rongalite-potash method of printing with cubes.
with dyes.

This method involves rapid recovery of the cube dye not as part of the printing ink, but directly on the fiber during processing of the fabric in a steamed recovery milling machine. In this case, however, special release forms of cube dyes in the form of printing pastes are required.

Progress of the work:

The composition of the printing ink for reproduction of the pattern includes, g/50 g:
Cube dye (paste) - 1.0
Glycerin - 0.5
Potassium carbonate (potash) 1:1 with water - 10.0
Rongalite 1:1 with thickening - 10.0
Thickener - up to 50 g
In order to save rongalite and intensify the printing process it is possible to use printing ink containing, g/50 g:
Cube dye (paste) - 1.0
Glycerin - 0.5
Potassium carbonate (potash) 1:1 with water - 10.0
Rongalite 1:1 with thickening - 5.0
Sodium sulfite 1:1 with water - 5.0

Urea - 4.0
Catalyst (dimethylglyoxime) - 0.1
Water - 5.0
Thickener - up to 50 g

The order of preparation of printing ink. Mix the dye with glycerine, then, each time mixing thoroughly, introduce in the following order: potash with water, rongalite with thickener and finally thickener and water to the required consistency. When using the second recipe, sodium sulfite (Na SO$_{23}$), previously dissolved in water, enter together with potassium carbonate and urea. Add the catalyst to the printing ink as an aqueous solution just before use. The printing composition is applied to the fabric using a grid template, dried with hot air in a desiccator and steamed in a laboratory burner at a temperature of 100° C for 15 minutes (when using the printing ink prepared according to recipe 1) or 3 minutes (for the second printing composition). After steaming, the color of the dye changes due to the transition of the dye into a soluble form. To obtain the original color, oxidize the dye on the fabric in a solution of potassium bichromate or hydrogen peroxide for 3-5 minutes. Finally, the sample should be washed with cold water, treated in a detergent solution at boiling for 5 minutes, rinsed with water and dried.

Assignment 2: Study the technology of printing cotton fabrics with cubes. with dyes using a two-step process.

When applying patterns using this method, it is recommended to use cube dyes with D mark (for suspension dyeing) or printing pastes. This technology imitates the suspension dyeing process.

Progress of the work:
Prepare printing ink of the following composition, g/50 g, for drawing:
Cube dye - 1.0
Thickener - up to 50 g
The printing composition is applied to the fabric using a mesh template, dried and processed in 100 ml of developer solution, which is an alkaline reducing agent solution containing, g/l:
Sodium dithionite (hydrosulfite) - 40.0
Sodium hydroxide 30% sodium hydroxide - 50.0
Sodium sulfate - 20.0
Water - up to 1000 ml
The temperature of the solution is 20° C, treatment time is 5 minutes. The sample is squeezed out on the laboratory plusovka and steamed in the laboratory burner at a temperature of 100° C for 2 minutes (until the color of the original dye changes). Oxidative treatment of the fabric after steaming and washing should be carried out as given in task 1.

Laboratory work No. 3
Printing of cellulose fiber fabrics with cubosols

Materials and chemicals required:
Cotton cloth 10x10 cm - 6 pcs.
Mesh template - 1 pc
Rubberized doctor blade - 1 pc
Drying cabinet - 1 pc

Laboratory drill press - 1 pc
Cubozole - 10 g
Glycerin - 30 g
Sodium carbonate - 50 g
Sodium nitrite - 25 g
Potassium chlorate ($KClO_3$) - 20 g
Ammonium rhodanide ($NH_4 CNS$) - 50 g
Sodium vanadate 1:1000 with water - 50 ml
25% ammonia - 50 ml
Thickening - 200 g
Surfactant - 100 g.

In practice, two methods of printing fabrics with cubozols are used:
1) Nitrite;
2) steamy.

In the printing process, cubozols combine well, allowing for a much wider range of colors, and they are also used to print products made from a blend of cellulose fibers and polyester fibers. This provides the advantages of cubozoles over water insoluble cube dyes.

Assignment 1. Study the technology of printing fabrics with nitrite cubesols. in a way.

This printing method eliminates the operation of fabric processing in the steam burner. Sodium nitrite is introduced into the printing ink as an oxidizing agent, and sodium carbonate is added to increase the stability of the composition. After printing and drying, the fabric is treated in a solution of sulfuric acid, providing hydrolysis of cubozene to leukocompounds.

Progress of the work:
Prepare printing ink of the following composition, g/50 g:
Kubozol - 1.0
Water (20° C) - 5.0
Sodium carbonate - 0.3
Sodium nitrite - 0.5
Thickener - up to 50 g

Dissolve cubozene in warm water and mix with thickener. Dissolve sodium carbonate and sodium nitrite separately in 5 ml of water and add to the printing ink while stirring. The printing composition is applied to the fabric using a grid template, dried and treated in 50 ml of sulfuric acid with a concentration of 30 g / l at a temperature of 60-70° C until the appearance of color. Drain the acid, wash the sample with warm water, hot soap solution (70-80° C) for 5 minutes, again with hot, cold water and dry.

Assignment 2. Study the technology of printing fabrics with cubosols by the steamer
in a way.

In this method of printing, color development takes place under steaming conditions. Potassium chlorate or sodium chlorate are used as oxidizing agents. To create an acidic environment, a potentially acidic reagent, ammonium rhodanide, is introduced into the printing ink. Hydrolysis and oxidation of cubosol to insoluble cubic dye occur only in the oxidizing ripper, where chlorates decompose with the release of oxygen, and ammonium rhodanide is hydrolyzed, creating an acidic environment.

Progress of work

Prepare printing ink according to the following recipe, g/50g:

Kubozol - 1.0

Glycerin - 3.0

Water - 10.0

Potassium chlorate ($KClO_3$) - 1.0

Ammonium rhodanide (NH_4 CNS) - 1.5

Sodium vanadate 1:1000 with water - 1.0

25% ammonia - 1.0

Thickener - up to 50 g

The order of preparation of printing dye: rub cubozol with glycerin and dissolve in water, add thickener and with constant stirring successively introduce ammonia, potassium chlorate (1:1 with water), ammonium rhodanide (1:1 with water), and before using the catalyst - sodium vanadate. The printed composition is applied to the fabric using a mesh template, dried and steamed in a laboratory burner for 5 minutes until the color of the cube dye appears. Wash the sample with warm water, a bitter detergent solution, again with water and dry.

Laboratory work No. 4
Printing of fabrics made of synthetic fibers and acetate silk with dispersed materials
dyes

Materials and chemicals required:

Fabric made of polyamide fibers 10x10 cm - 6 pcs.

Fabric made of polyester fibers 10x10 cm - 6 pcs.

Mesh template - 1 pc

Rubberized doctor blade - 1 pc

Drying cabinet - 1 pc

Laboratory drill press - 1 pc

Disperse dye - 25 g

Glycerin - 25 ml

Urea - 100 g

Stearox - 20 g

Alginate thickener - 200 g

Thickener - 500 g

Surfactant - 200 g

Dispersed dyes are widely used in printing textile products made of acetate, polyamide and polyester fibers, which have the property of thermoplasticity. Therefore, in all methods of patterning on fabric with the help of disperse dyes, the processing of textile materials at a temperature that is 40-50° C higher than the glass transition temperature of these fibers is provided. Disperse dyes are widely used in printing on linen knitted fabric. They provide bright and saturated colors, well combined, which allows to significantly expand the color range. In practice, two technological modes of printing are used:

1) thermal;

2) method of transfer thermal printing (sublimation).

Assignment 1: Study the technology of thermal method of textile printing. with dispersed dyes.

Progress of the work:
Prepare printing ink according to the following recipe, g/50 g:
Dispersed dye - 0.5
Surfactant (50 g/l) - 2.0
Glycerin - 0.5
Urea - 0.5
Thickener - up to 50 g
The following thickeners are recommended for printing with disperse dyes: tragant, alginate, CMC, solvitose.

The samples printed with the help of mesh templates should be dried at a temperature of 90-100° C and subjected to heat treatment for dye fixation. Products from acetate and polyamide fibers are heat fixed at a temperature of 101-105° C in saturated water vapor for 20-30 minutes. Samples from polyester fibers, which practically do not swell in steam environment, after printing and drying are treated with dry hot air for 1 minute at a temperature of 190° C. After thermofixation, the samples should be washed in warm and cold water, in hot surfactant solution for 5-7 minutes, finally in warm water and dried.

Assignment 2: Study the technology of transfer thermal printing on fabrics made of
   thermoplastic fibers.

The peculiarity of disperse dyes, which distinguishes them from dyes of other classes, is the ability at certain temperatures (usually above 130-140° C) to ignite (sublimate), i.e., bypassing the liquid phase, pass into the gaseous state, forming saturated vapors. This property of dispersed dyes is the basis for the method of transfer thermal printing, the essence of which is that the pattern is first applied to a paper substrate carrier, and then from it is transferred to the textile product by close contact with it and exposure to high temperature. The most important advantage of this method is the exclusion of labor- and material-intensive operations - washing and subsequent drying. Transfer thermal printing allows you to reproduce on the fabric up to 100 different colors and shades, while achieving high clarity of the pattern.

Progress of the work:
Prepare the backing paper on which to print the pattern using grid templates or by hand. Use the following composition of printing ink, g/50 g:
Dispersed dye - 2.0
Urea - 5.0
Water 40-50° C - 15,0
Stearox - 2.0
Alginate thickener - up to 50 g

Printing should be carried out as follows. On a sample of pre-prepared (subjected to thermal stabilization) web put the paper (face side) and place under the iron, tightly pressing its heated surface to the wrong side of the paper. The temperature of thermofixation 180-200° C, duration 30-40 s. During the pressing process, the dye sublimates from the backing paper, reproducing the pattern on the fabric. After printing, neither steaming to fix the dye nor washing is required.

Laboratory work No. 5
Printing fabrics with pigments

Materials and chemicals required:
Cotton cloth 10x10 cm - 6 pcs.
Fabric made of polyamide fibers 10x10 cm - 6 pcs.
Fabric made of polyester fibers 10x10 cm - 6 pcs.
Mesh template - 1 pc
Rubberized doctor blade - 1 pc
Drying cabinet - 1 pc
Bronze powder - 25 g
Polyvinyl acetate emulsion - 100 g
Metazine - 50 g
Ammonium chloride - 25 g
Emucryl M - 25 g
CMC thickener - 400 g
Emulsion thickener - 100 g
Polyvinyl acetate emulsion - 100 g

Pigments have no affinity for any of the fibers, so they are fixed to the textile material with special binders. Pigments are mainly used in printing and less frequently in dyeing. They have a high dyeing power and the dyes are characterized by high resistance to wet processing and to light. These dyes can be applied to any fabric, regardless of its chemical nature, both white and dyed, because their fixation occurs with the help of binders - substances capable of wetting the fiber well when applying a dyeing solution or printing ink, and when heat treatment to form on the surface of the fabric insoluble in water transparent, elastic film, firmly bonded to the fiber material. In the process of pigment printing special requirements are imposed on the thickeners of printing inks: they must be removed from the fabric during high-temperature treatments (drying, thermofixation), as the pigment printing technology excludes washing and subsequent drying operations. The disadvantage of pigment printing is the low resistance of dyeing to friction, because the binder film fixing the dye on the fiber is distributed in the upper layers of the fiber.

Task 1: Study the technology of printing fabrics with mineral pigments (metal powders, carbon black, zinc or titanium oxides, insoluble in water colored inorganic salts).

The technological process of printing fabrics with pigments includes the following operations: preparation of printing ink, application of printing composition on fabric, drying of fabric and its thermal treatment with hot air at a temperature of 140-160° C for 3-5 minutes. Printing can be carried out both on white and dyed fabrics. The CMC thickener can be replaced with acrylic or emulsion thickener.

Progress of the work:
On the dyed or white (colored pigments) fabric apply with the help of a grid template the printing composition containing, g/100g:
Bronze powder - 4.0

Polyvinyl acetate emulsion - 15.0
Metazine - 15.0
Ammonium chloride - 2.0
CMC thickener - up to 100 g

Prepare the printing ink by adding all components in sequence. Introduce the catalyst (ammonium chloride) into the printing ink immediately before use. Polyvinyl acetate emulsion and methazine act as binders. The printing compound is applied to the fabric using a mesh template, dried at 100° C, then heat-treated at $160\text{-}170^\circ$ C for 3 minutes.

Task 2: Study the effect of thickener on the quality of pigment printing.

Progress of the work:
Prepare two printing compositions containing, g/100g:

Composition 1		Composition 2	
Pigment	- 2,0	Pigment	- 2,0
Polyvinyl acetate emulsion	- 15,0	Polyvinyl acetate emulsion	- 15,0
Metazine	- 15,0	Metazine	- 15,0
Ammonium chloride	- 2,0	Ammonium chloride	- 2,0
Starch thickener	- 66,0	Emulsion thickener	- 66,0

Assignment 3: Study the influence of the nature of the binder composition on the quality of printing
pigments.

Progress of the work:
Prepare two printing compositions containing, g/100g:

Composition 1		Composition 2	
Pigment	- 2,0	Pigment	- 2,0
Polyvinyl acetate emulsion (PVA)	- 15,0	Emucryl M	- 30,0
Metazine	- 15,0	CMC thickener	- 68,0
Ammonium chloride	- 2,0		
CMC thickener	- 66,0		

Print two samples of cotton fabric with the given compositions, dry at 100° C and heat-treat for 3 minutes at 160° C. Test for resistance to washing and abrasion.

Laboratory work No. 6
Printing fabrics with direct dyes

Materials and chemicals required:
Cotton cloth 10x10 cm - 6 pcs.
Mesh template - 1 pc
Rubberized doctor blade - 1 pc
Drying cabinet - 1 pc
Laboratory drill press - 1 pc

Iron - 1 pc
Direct dye - 10 g
Urea - 50 g
30% formaldehyde - 50 ml
Dicyandiamide (DCU) - 50 g
Carbomol - 50 g
Metazine - 50 g
Triethanolamine - 20 ml
Ammonium chloride - 50 g
Thickening - 200 g
Starch thickener - 100 g
Rongalite - 50 g
Cube dyes - 10 g

Direct dyes are used relatively rarely in fabric printing because of the coloring instability to washing. However, these dyes are cheap, have a full range of colors and uncomplicated technology of reproduction of patterns. Therefore, when creating conditions that ensure the fixation of dyes on the fiber, their use in printing is of interest.

Assignment 1: Study the process of printing fabrics with direct dyes by the method of
then attaching them to the fiber.

Progress of the work:
Prepare printing ink according to the following recipe, g/100 g:
Direct dye - 2.0
Hot water - 10.0
Thickening - up to 100 g
In the process of preparation the dye is dissolved in hot water (70-80° C) and mixed with cold starch thickener. After printing, dry the fabric sample, steam it for 5-7 minutes in saturated water vapor and process it in 50 ml of dicyandiamide DCU solution for 15 minutes at 80° C. The fabric is pressed on a plusovka in a straightened form and dried with an iron.

Task 2. Print the fabric with direct dyes with simultaneous color fixation.

Prepare printing ink according to the following recipe, g/100g:
Direct dye - 2.0
Hot water - 10.0
Urea - 2.0
30% formaldehyde - 5.0
Dicyandiamide (DCU) - 5.0
Triethanolamine - 2.0
Ammonium chloride - 0.5
Thickening - up to 100 g

Dissolve dye in hot water, add thickener, mix thoroughly and add triethanolamine, formaldehyde, urea, dicyandiamide and ammonium chloride sequentially with constant stirring. Instead of formaldehyde and dicyandiamide it is possible to use precondensates of thermosetting resins (carbomol, otexide, metazine, etc.). The prepared printing compound is applied to the fabric using a mesh template. Then the sample is dried, subjected to heat treatment for 3-5 minutes at a temperature of 150-170° C, washed and dried.

Assignment 3: To test the colorfastness to washing of fabrics printed with with direct dyes.

Progress of the work:
Prepare a sample printed without color fixation. For this purpose, apply a primer pattern on cotton cloth with a composition containing, g/100 g:
Direct dye - 2.0
Hot water - 10.0
Starch thickener - up to 100 g 69

Dry the sample, steam for 5-7 minutes over boiling water, rinse with warm water and dry. Cut three samples of fabric, printed according to the methods of tasks 1, 2, 3, each into two parts. Leave one part for comparison, sew a white cloth under the second part and place the samples in 30 ml of a solution containing 3 g/l of soap and 3 g/l of soda (each in a separate beaker), heat the solutions to 50° C and keep the fabrics in them for 30 minutes, maintaining the specified temperature. Then the samples were washed with warm water and dried. The durability of colors is assessed visually by staining of white fabric and the background of the pattern.

Coloring of fabrics by the method of etching printing. Etching printing provides obtaining white and colored patterns on the dyed fabric by applying to it compositions that destroy the color of the main background. For color etching use dyes, in direct printing, which are introduced into the printing ink substances capable of etching the color of the main background. The greatest use is found reducing etchings, in which the substance that destroys the color of the background, serves as a reducing agent - rongalite. To reproduce colored patterns in this case, cube dyes are used.

Laboratory work No. 7
Obtaining white and colored etchings on direct dyed fabrics
dyes

__Materials and chemicals required:__
Bleached cotton cloth 10x10 cm - 6 pcs.
Mesh template - 1 pc
Rubberized doctor blade - 1 pc
Drying cabinet - 1 pc
Laboratory drill press - 1 pc
Iron - 1 pc
Direct dye - 10 g
Sodium chloride - 30 g
Sodium carbonate - 25 g
Rongalite - 200 g
Potassium carbonate (potash) - 100 g
Direct dye thickener - 400 g
Cube dye - 20 g
Cubozole - 10 g
Glycerin - 20 g
Potassium chlorate ($KClO_3$) - 20 ml

Ammonium rhodanide (NH_4 CNS) - 20 g
Ammonium vanadate - 20 g
Cubosol thickener - 100 g
30% chrompik solution (K_2 Cr O_{27}) - 100 ml

Task 1. Obtain a white pattern on fabric dyed with direct dyes.

Progress of work
Dyeing. Dye bleached fabric at a temperature of 90° C for 20 minutes in 100 ml of solution containing, g/l:
Direct dye - 1.0
Sodium chloride - 3.0
Sodium carbonate - 0.5
Water - up to 1000 ml 70
Rinse the stained sample with warm water and dry.
Preparation of printing composition. To obtain white etching prepare 20 g of printing ink of the following composition, g/100g:
Rongalite - 20.0
Potassium carbonate (potash) - 10.0
Water - 10.0
Thickening - up to 100 g
Rongalite should be ground in a porcelain mortar and dissolved in water. Add the necessary amount of potash to the solution, add thickener, mix to a homogeneous mass and keep for 10-15 minutes until complete dissolution of rongalite.
Fabric printing. Apply the printing compound on the dyed fabric with the help of a mesh template. Dry and steam the sample for 5-10 minutes. For processing in a steam environment, place a sheet of filter paper corresponding to the size of the sample on the front side of the printed fabric and roll it into a tube with the paper inside. After the white pattern is revealed, rinse the fabric with warm water and dry it.

Task 2. Obtain a colored pattern with cube dyes on fabric dyed with with direct dyes.

The use of cube dyes for color reduction etching is based on their ability to change into a soluble form when reduced in an alkaline environment. The composition of cube printing inks is the same as for direct printing on white fabric, but the amount of rongalite is increased. This is necessary because the reducing agent performs dual functions:
- discolors the background coloring;
- reduces the cube dye to leukoic acid.
Progress of the work:
Dyeing. Dye a sample of cotton fabric with direct dyes according to the technology of task 1.
Preparation of the printing composition. Printing ink contains, g/100 g:
Cube dye - 10.0
Rongalite - 30.0
Potassium carbonate (potash) - 10.0
Water - 10.0 71
Thickening - up to 100 g

Prepare 20 g of printing ink: dissolve rhongalite and potash in water, add thickener, mix and add dye. Keep the prepared composition for 5-10 minutes until the components are completely dissolved, mix and use for printing.

Fabric printing. Apply a colored pattern on the dyed fabric using a grid pattern and dry it. After drying, cover the sample with a sheet of filter paper (see task 1) and steam it for 5-20 minutes until the color of the dye changes (its transition into a soluble form). Drops of condensate should not get on the sample, otherwise the coloring will be blurred. After steaming, rinse the fabric with warm water, treat with chrompik solution ($K_2 Cr O_{27}$) and again with water, then dry.

Task 3. Get a colored pattern cubozolami on the fabric dyed with straight lines

dyes by oxidative etching.

In this type of printing, the oxidizing agent - potassium or sodium chlorate - plays the role of etching agent. At the same time, the oxidizing agent converts the cubozene into insoluble cube dye. The manifestation of the coloring takes place under steaming conditions. To create an acidic environment necessary for the hydrolysis of cubozene, a potentially acidic reagent - ammonium rhodanide - is introduced into the printing composition, and vanadium salts are used as a catalyst for the oxidation process.

Progress of the work:

Preparation of printing composition. Prepare 20 g of printing ink containing, g/100g:

Kubozol - 1.0

Glycerin - 1.0

Water - 20.0

Potassium chlorate ($KClO_3$) - 2.0

Ammonium rhodanide ($NH_4 CNS$) - 2.0

Ammonium vanadate - 1.0

Thickening - up to 100 g

Mix cubazole with glycerine and dissolve in half the amount of water. Dissolve potassium chlorate, ammonium rhodanide and ammonium vanadate in the remaining amount of water, add thickener and stir thoroughly. Combine the obtained composition with the cubozolium solution and keep until the components are completely dissolved.

Fabric printing. Prepared printing composition with the help of a grid template to apply to the fabric, colored with a direct dye (in a light tone). Dry the sample, process in a steam environment for 10-30 minutes until background discoloration and cubosol coloration, rinse with warm water and dry.

Coloring of fabrics by back-up printing methods. Backup printing methods involve dyeing fabrics after the printing ink has been applied. In this way you can get white and colored patterns on dyed fabrics without etching the main background, as the reserve composition of printed ink prevents the formation and fixation on the fabric dye used to paint the background. A distinction is made between mechanical and chemical reserves. At mechanical method of reservation the printed composition forms a film on the fabric impermeable to the dye solution (like batik-process). Chemical reserves prevent the formation of coloration in the places of drawing due to the chemical interaction of the reserving substance with the components of the dyeing solution.

Laboratory work No. 8
Creating various artistic effects on fabrics using the printing method

Bleached cotton cloth 10x10 cm - 6 pcs.
Fabric made of polyamide fibers 10x10 cm - 6 pcs.
Fabric made of polyester fibers 10x10 cm - 6 pcs.
Viscose-polyester fabric - 6 pieces
Viscose-polyamide fabric - 6 pieces
Rubberized doctor blade - 1 pc
Drying cabinet - 1 pc
Laboratory drill press - 1 pc
Active coloring agent - 10 g
Urea - 50 g
Soda - 20 g
Thickening agent for active coloring - 400 g
White spirit - 40 ml
Emulsifier (surfactant) - 20 g
Disperse dye - 50 g
Emulsion thickener - 400 g
Titanium oxide - 50 g
Dibutyl phthalate - 50 g
Metazine - 100 g
Ammonia - 20 ml
Polyvinyl alcohol - 100 g
CMC thickener - 500 g
Alginate thickener - 400 g
Ultramarine for undercolor - 25 g
Active coloring agent - 15 g
Sodium hydrogen carbonate - 25 g
Disperse dye - 10 g
Glycerin - 50 g
Resorcinol - 50 g
Thickening agent for disperse coloring - 400 g
Sulfuric acid - 100 ml
Acid-resistant thickener - 200 g
Soda - 100 g

Coloring by printing method allows to obtain highly artistic textile products by reproducing on them various effects imitating watercolor patterns, lace, matte shade, etc.

Task 1. Obtain the effect of watercolor printing on cotton fabric.

The essence of the method is that the drawing on the fabric does not form a clear contour, but has a vague shape. The colors can be overlapped, creating a variety of complex tones and halftones, like a watercolor drawing on paper.

Progress of the work:

Preparation of the printing compound. Prepare two printing compositions based on active dyes of different colors. The ink should be more fluid than in conventional printing. Composition of printing ink, g/100 g:

Active colorant - 1.0
Water - 50.0

Urea - 2.0
Soda - 2.0
Thickening - up to 100 g

Printing of fabric. Soak a sample of cotton fabric in water, squeeze it out on a plusovka and apply on it (without drying) a two-color pattern with two templates, partially layering one color on another. Steam the cloth in a flask for 3 minutes, rinse and dry.

Task 2. Obtain the effect of watercolor printing on fabrics made of acetate or polyamide fibers.

The watercolor effect in this case is achieved by using emulsion thickeners.
Progress of the work:
Preparation of printing composition. Prepare emulsion thickener containing, g/100 g:
White spirit - 4.0
Emulsifier (surfactant) - 2.0
Water - up to 100 g

Stir the resulting mixture intensively to obtain a viscous system in which the oil droplets are evenly distributed in the aqueous medium and use it to thicken the printing ink. Prepare the printing compound according to the following recipe, g/100 g:
Dispersed dye - 10.0
Water ($60°$ C) - 10.0
SURFACTANT - 1.0
Urea - 2.0
Emulsion thickener - up to 100 g

Printing of fabric. Print a sample of fabric made of capron or acetate silk with the prepared printing compound, dry, steam for 15-30 minutes in water vapor, rinse with warm water and dry.

Task 3. To obtain the effect of matte whitewash on the fabric.

In this case, a pigment printing method is realized, wherein titanium oxide is used as a white pigment.
Progress of the work:
Preparation of the printing compound. Prepare 20 g of printing compound containing, g/100g:
Titanium oxide - 25.0
Dibutyl phthalate - 10.0
Metazine - 20.0
Urea - 1.0
Ammonia - 2.0
Polyvinyl alcohol - 10.0
Water - 7.0
CMC thickener - up to 100 g

Mash titanium oxide with dibutyl phthalate, add polyvinyl alcohol, ammonia and water. Mix metazine with thickener, add urea. Combine both compositions, mix thoroughly and use to print the dyed fabric.

Printing of fabric. Apply the prepared composition on the dyed fabric, dry the sample, steam it for 15 minutes in saturated water vapor, rinse and dry.

Task 4: Get a halftone printing effect on fabric.

To obtain this effect it is advisable to use cube or active dyes. The essence of the method is that the printing ink is applied to the areas of the fabric previously printed with thickeners. This achieves dilution of the printing composition and brightening of the pattern.

Progress of the work:

Preparation of printing composition. Prepare 20 g of clarifying composition containing, g/100g:

Alginate thickener - 50.0

Ultramarine for undercolor - 0.5

Water - 49.5

Prepare 20 g of printing ink based on active dye containing, g/100 g:

Active colorant - 3.0

Urea - 7.0

Sodium hydrogen carbonate - 5.0

Water - 15.0

Alginate or CMC thickener - up to 100 g

Dissolve urea in water, add active coloring, mix with the thickener and lastly introduce sodium bicarbonate.

Fabric Printing. On the cotton fabric with the help of a grid pattern apply the brightening compound and dry the sample. Then apply printing ink on the same places (the pattern can be shifted) and after drying, the sample is steamed for 10 minutes, rinsed and finally dried. Observe the effect obtained by comparing the intensity of coloration in the pre-brightened areas and in the unprotected areas.

Task 5. Obtain a shrunken effect on a fabric made of polyamide fiber.

Obtaining the reaped effect is based on the swelling of polyamide fibers under the action of organic solvents (resorcinol).

Progress of the work:

Preparation of printing composition. Prepare 20 g of printing ink containing, g/100 g:

Dispersed dye - 1.0

Glycerin - 1.0

Resorcin - 10.0

Thickening - up to 100 g 91

Fabric Printing. Apply the pattern on the kapron cloth using a template with a small pattern. Dry the pattern, steam it for 15 minutes, treat it in 50 ml of baking soda and soap solution for 5 minutes, rinse and dry it.

Assignment 6. Study the possibility of combining the processes of production of reaper
effect and halftone printing.

In this case, the effect of organic solvent printing is combined with the dyeing process. The coloring is more intense in the areas where the pattern is applied compared to the background coloring.

Progress of the work:

Preparation of printing composition. Prepare 20 g of printing ink containing, g/100 g:

Resorcin - 20.0
Water - 10.0
Thickening up to 100 g
Preparation of the dye solution. Prepare 100 ml of dye solution containing, g/l:
Dispersed dye - 5.0
SURFACTANT - 5.0
Water - up to 1000 ml

Fabric printing. On a white kapron cloth, apply the printed composition for the printed effect in the form of a pattern, dry it and steam it for 15 minutes.

Dyeing. Dye the printed sample in the prepared dye solution for 15 minutes at boiling temperature. Then rinse and dry the fabric. Explain the formation of more intense coloration where the printing compound is applied.

Task 7. Get a transparent pattern on the fabric, imitating lace (method scorching).

The essence of the method is that fabrics made of a mixture of hydrated cellulose and synthetic fibers, resistant to the action of acids, are printed with compositions containing acid or salts releasing it during hydrolysis. After printing and drying the fabric is subjected to heat treatment with hot air, in which the cellulose component is hydrolyzed and washed off by washing. A transparent pattern is produced on the fabric.

Progress of the work:
Preparation of printing composition. Prepare printing ink containing, g/100 g:
Sulfuric acid - 2.0
Acid-resistant thickener - 98.0

Printing fabric. For printing use viscose-polyester or viscose-polyamide fabric. Apply the pattern on the sample with the prepared printing ink, dry the fabric, heat treatment with hot air for 5-10 minutes at 120° C or 2 minutes at 140° C, rinse with water, treat in a solution of soda (1 g/l) to neutralize the acid, finally rinse and dry.

Literature

1. Khudova L. N. On the current situation in the light industry in the Republic of Kazakhstan // Innovative technologies of goods production, improving the quality and safety of light industry products: Proceedings of the International scientific-practical conference, Almaty, May 25, 2012. - C. 59-61.
2. Program for the development of light industry in the Republic of Kazakhstan for 2010-2014. - Astana, 2010.
3. Krichevsky, G.E. Chemical technology of textile materials: textbook for universities in 3 vol. Vol. 2 / G.E. Krichevskiy. - Moscow, 2000.- 540 p.
4. Zhuravleva, N.V. Coloring of textile materials: textbook for universities / N.V. Zhuravleva, M.V. Konovalova, M.A. Kulikova. - Moscow: Kosygin Moscow State Technical University, 2007. -368 c.
5. Kiselev, A.M. Fundamentals of foam technology of textile materials finishing / A.M.Kiselev. - SPb: SPGUTD, 2003. - 551 c.
6. Finishing of cotton fabrics: reference book / edited by B.N. Melnikov. - Ivanovo: publishing house "Talka", 2003. - 484 c.
1. Heat Press Manual Model: ECH - 800 Manuall-INSTRUCTION.
2. Video tutorial: ECH-800 8 in 1 Combo Heat Press Machine for T-shirt,Plates (2 sizes), Mugs(4 sizes), Caps - YouTube [360p].
7. Makachev, A.N. Computer technologies in textile industry / A.N. Makachev // Textile chemistry. - 2004.-Special. issue No. 4.-C. 20-22.
8. Volkhonskaya, N.S. Velvet effects when printing textile materials by electroflocking / N.S. Volkhonskaya, T.A. Dergacheva // Textile chemistry. - 2003.-Special. issue No.1.- C. 69-71.
9. Kovsh, I.S. INKJET - revolutionary direction in textile printing / I.S. Kovsh // Textile chemistry. - 2003.-Special. issue No.2.-C.12-13.
10. Stepanova, E.I. Problems and prospects of transfer printing on textile /E.I.Stepanova // Textile chemistry. - 2003.-Special. issue No.2.-C.14-15.
11. Advertising brochures of the following companies: Stork, Riggiani, Unica, CromoTex, Huntsman (Ciba), DyStar, Clariant, BASF, Bezema, Minerva, Sinthesia, NPO KRATA, Kisco, NPO KATION.
12. Kharkharov A.A. Printing and final finishing of fiber materials. Textbook.-L.:Izd. of Leningr. un-sta, 1984.-129 pp.
13. Lobanova L.A. Dyeing, printing and painting of textile materials / Manual. - M.:2013.- p.608
14. Badanov K.I. Activation of chemical-textile processes of finishing production. Taraz: Taraz University, 2013. - 223 c.
15. Safonov V.V. Intensification of chemical-textile processes of finishing production. Study guide. -M., MSTU named after A.N. Kosygin, 2006. -405c.
16. Sadova S.F. et al. Ecological problems of finishing production. Textbook for universities. Edited by S.F. Sadova. - M.: RIO MGTU, 2002. - 284 c.

I want morebooks!

Buy your books fast and straightforward online - at one of world's fastest growing online book stores! Environmentally sound due to Print-on-Demand technologies.

Buy your books online at
www.morebooks.shop

Kaufen Sie Ihre Bücher schnell und unkompliziert online – auf einer der am schnellsten wachsenden Buchhandelsplattformen weltweit! Dank Print-On-Demand umwelt- und ressourcenschonend produzi ert.

Bücher schneller online kaufen
www.morebooks.shop

info@omniscriptum.com
www.omniscriptum.com

Printed by Books on Demand GmbH, Norderstedt / Germany